TRADE, INDUSTRIAL CO-OPERATION AND TECHNOLOGY TRANSFER

To our families

Trade, Industrial Co-operation and Technology Transfer

Continuity and change in a new era of East-West relations

MALCOLM R. HILL
CAROLINE M. HAY
Loughborough University Business School

Avebury
Aldershot · Brookfield USA · Hong Kong · Singapore · Sydney

Published by
Avebury
Ashgate Publishing Limited
Gower House
Croft Road
Aldershot
Hants GU11 3HR
England

Ashgate Publishing Company
Old Post Road
Brookfield
Vermont 05036
USA

British Library Cataloguing in Publication Data

Hill, Malcolm R.
Trade, Industrial Cooperation and Technology Transfer: Continuity and Change in a New Era of East-West Relations
I. Title II. Hay, Caroline M.
338.9260947

ISBN 1 85628 502 2

Printed and Bound in Great Britain by
Athenaeum Press Ltd, Newcastle upon Tyne.

Contents

Foreword

Carl McMillan

'New eras' in East-West economic relations have been proclaimed prematurely before. Certainly relations between the industrialized economies of the capitalist West and those of the heretofore socialist East remain controversial and risky, and plagued with many of the problems of the past. It is clear, nonetheless, that the revolutions of 1989 in Eastern Europe and the disintegration of the Soviet Union in 1990-91 have dramatically altered the context and much of the content of these relations in the 1990s.

The effects of the recent upheavals on East-West relations have been diverse and contradictory, however. On the one hand, the dismantling of the traditional institutions of central planning and the collapse of the Comecon regional system have served dramatically to open up the Eastern economies to relations with the rest of the world and with Western Europe in particular. They have created a whole range of new opportunities for Western firms seeking to expand in the area. These opportunities have been reinforced by the general improvement in the post-Cold War political climate of relations and by the programmes of Western economic and technical assistance to the Eastern countries that have replaced earlier embargoes and sanctions.

At the same time, however, the deep economic (and political) crises that have accompanied political and institutional change in all the countries of the area in the early 1990s have served to limit the immediate possibilities for the expansion of East-West commerce. Growing economic and political instability has substantially raised the risks to Western firms operating in the area in these early years of transition.

However sweeping change in the area has appeared to be, the former socialist economies and societies are not a *tabula rasa*. The legacies of the past constrain and shape the processes of their domestic transformation and their reintegration into the world economy. The course of current events can therefore only be understood in the context of what has preceded them.

The volume at hand is highly useful in delineating the diverse, and often contradictory, conditions that Western firms currently face in pursuing business strategies in the areas. Its case study approach is particularly valuable at a time when the validity of macroeconomic data and the projections based upon them must be treated with great caution.

One of the authors has been closely following the subjects covered here for a number of years. He brings to the study of current developments a deep historical perspective. In fact, the present volume specifically follows up research carried out on many of the same British firms in an earlier period. It is thus able to bring out sharply, and to illustrate with practical examples, the elements of both continuity and change in the complex, current situation.

This volume examines the themes of trade, industrial cooperation and technology transfer in an indisputably new era in the history of East-West relations. The information it assembles and the expertise and perspective which the authors bring to its interpretation will be of direct relevance both to firms with prior experience in conducting business in the area, and to the many newcomer firms that have been attracted to the opportunities that have opened up.

Carl McMillan
Ottawa, 1993

Acknowledgements

The authors wish to record their thanks to the company executives who agreed to take part in this study, providing time for interviews and subsequent discussion of the initial case study drafts. The authors are also grateful to Mr L J Rogers for assistance in the preparation of Chapter 4 and to Mr R Hamilton for his comments on an earlier draft of Chapter 8.

The funding for Miss Hay's research was provided by a grant administered by the Loughborough University Research Committee. Professor Hill's research has been supported by funding from the Nuffield Foundation Social Sciences Small Grants Scheme. The authors wish to record their thanks for financial support from those two institutions.

The authors' University Department provided the facilities to enable the research to be carried out and the manuscript to be prepared. Acknowledgements are consequently made to Professor J Sizer and Professor D A Buchanan who were the respective Directors of the Loughborough University Business School during the time period when this book was researched and written. Finally, the authors wish to acknowledge the patience and efficiency of Miss Lynne Atkinson during the completion of the final manuscript.

1 Introduction

Scope of the research

The research project described in this book has been an investigation into the current export trading activities of British companies in the former socialist countries of Eastern Europe (Bulgaria, Czechoslovakia, German Democratic Republic [GDR], Hungary, Poland, Romania and the Union of Soviet Socialist Republics [USSR]) during a period of recent rapid political and economic change. In addition, the research has also included a study of the commercial activities of British companies which are owned by business organizations based in the former socialist countries of Eastern Europe.

The methodology used in this project has been one of follow-up case study research of a sample of some forty-five companies, which had previously been studied by one of the present authors some ten to fifteen years previously.[1] The major changes in the political and economic environment that have taken place in the former socialist countries since those previous studies, are described in the following section of this chapter.

Political and economic context

The geographical region occupied by the former socialist countries has undergone fundamental political change since 1987, and much of that change has caused consequent economic effects. The majority of these changes were directly linked to Mikhail Gorbachev's appointment in 1985 as General Secretary of the Communist Party of the Soviet Union, and the policies of open discussion (*glasnost*) and economic restructuring (*perestroika*) introduced following that appointment.[2] The direct consequence of glasnost was the more open nomination of candidates for election to the Soviet all-Union and Republican parliaments in 1989 and 1990, in which many of the more traditional Communist Party officials were rejected by the electorate, particularly in the republican parliaments.

This paved the way for the political leadership to propose a more radical programme of reform, particularly in the economic sphere.[3]

Although several different programmes were proposed for economic reform in the USSR in the late 1980s and early 1990s, the overall objective was that of a movement from a centrally planned to a market economy in order to improve the productivity of resources, increase the pace of product and process innovation, and match supplies more closely to demand. The differences in the proposed programmes related to the expected overall shape of the economy in terms of the proportions of publicly owned and privately owned resources, and the pace and defined stages of reform from the planned economy. Laws on the rights of industrial enterprises were introduced in 1987, to decrease the organizational dependence of the factory from its administratively superior industrial ministry, and to provide a greater level of responsibility for profitability and investment.[4] The organizational structure of foreign trade was also modified at approximately the same time, as some foreign trade organizations were transferred from the Ministry of Foreign Trade to relevant industrial ministries, responsible for co-ordinating the manufacture of those same products; and a number of individual factories were also granted the right to engage in import and export activities.[5] The aims of these organizational changes were to reduce the organizational barriers between producers and foreign traders, and create closer links between Soviet users and manufacturers, and foreign markets and sources of supply. At approximately the same time, laws were passed to establish a framework of legislation by which Western companies could invest in joint ventures with Soviet organizations. The general view prevailing in the West, therefore, was originally one of optimism in the late 1980s, as the Soviet Union was seen as casting away the previous restrictions to private enterprise, and becoming more open to Western business.

Gorbachev also made it clear to Western governments that he wished the atmosphere of mutual hostility between the West and the USSR to come to an end, and to reduce Soviet investment in armed forces and defence equipment production, in order to release resources to meet civilian demand. These proposals were welcomed by Western governments, and prompted discussion of how the West may assist the USSR in this process of 'conversion' from military to consumer goods production, and general economic development.[6]

The apparent changes in Soviet policy were also welcomed in Eastern Europe, where the post-1945 communist policies had never been widely accepted amongst the population as a whole. It was felt, therefore, that each country in Eastern Europe could follow its own process of reform without the fear of invasion from the USSR as suffered by Hungary in 1956 and Czechoslovakia in 1968.[7] The effect of these changes in Eastern Europe are now apparent with the communist parties being removed from power in Czechoslovakia, Hungary and Poland and the previous GDR becoming part of a unified German state. The most important consequences of these changes were the break-up of the military alliance with the USSR (the Warsaw Treaty Organization) and the cessation of bilateral trade between the socialist countries through the framework of the Council for Mutual Economic Assistance (CMEA or Comecon) in 1991.

Many of these countries have attempted to move rapidly from a planned to a market economy,[8] and to attract Western technology and investment to assist in that process.

During the late 1980s therefore, the Soviet Union and Eastern Europe were considered to be 'open for business', but various problems began to be identified not long afterwards.[9] In the USSR, for example, the changes which were introduced caused economic problems as supplies of many products in an economy which had always previously been constrained by shortages, became even more scarce. In addition, the objectives of self-sufficiency of industrial enterprises created concern over possible bankruptcies and subsequent unemployment. The associated food shortages led to fears of social disorder, with the subsequent coup attempt in August 1991 by some of the more conservative senior members of the government, army and Communist Party.

The power vacuum left by the failure of the coup was filled by the Russian republican government taking over many of the institutions of the previous Soviet government, and the assets of the Communist Party were seized by the Russian government in view of the Party's alleged implication in the attempted coup. The subsequent growth in republican power led to claims of ownership by each republic of the assets of the Soviet government within that republic. The corresponding loss in power of the central government led to the resignation of President Gorbachev, and the subsequent vote by the Soviet parliament for its own disestablishment in December 1991. The majority of the previous republics became members of a Commonwealth of Independent States (CIS), although some central control continued over the armed forces, and currency policy. It is within that context of political and economic uncertainty, particularly within the former USSR, that this present research has taken place.

Reference is consequently made throughout this book to the constituent states of the CIS, which occupy the same territory as the previously Soviet Socialist Republics constituting the federal Union of Soviet Socialist Republics. The most important of these republics in terms of industrial output and natural resources are Russia, Ukraine, Kazakhstan, Belarus (or Belorussia), and Azerbaijan. References are also made in this book to former Soviet Socialist Republics which chose not to become members of the CIS, namely Georgia and the Baltic states (Estonia, Latvia and Lithuania).

It is also important to note that 'Russia' or the 'Russian republic', to which the authors make frequent reference, is itself a federation comprising 'administrative areas' (*krai*), and 'autonomous republics' containing the local concentration of a particular national group. Since the establishment of Russia as a separate state, some of the autonomous republics have begun to seek increased decentralization within the federation.

Research methodology

As explained in a previous section of this chapter, the research project described in this monograph has been an investigation of the current export trading policies and practices of British companies towards the former socialist countries of Eastern Europe, during the recent period of accelerated political and economic change. There are clearly several ways in which this could have been analysed, including the study of trade statistics and the random sampling of companies engaged in exporting to that market. The approach used by the present authors, however, has been follow-up case study research using the structured interview method, of more than forty UK companies engaged in trade and technology transfer activities with the former Soviet Union and Eastern Europe. As already mentioned, the experiences of these companies in that market, from the late 1960s to the early 1980s, had been previously published by one of the present authors, and the follow-up studies were consequently chosen to provide an element of longitudinal research on company-level trading behaviour during a period of rapid market change.

These companies were also chosen for further research because of their well established experience and previous business commitment to the Soviet and Eastern European region, and the expectation that they would have the necessary expertise to modify their practices in such a manner that they could remain competitive in those markets during a time of accelerated change. In addition, the use of contacts established in these companies during the previous case study research was considered likely to yield information more quickly and cost effectively than the seeking of new contacts in different companies. The interviews were carried out between September 1991 and April 1992, covering the topics provided in the appendix, relating to each chapter.

The remainder of this book provides accounts of the companies' operations in case study format, divided into the following chapters according to the type of activity which had been carried out at the time of the previously published research:

- export marketing of capital goods, including machine tool technology transfer, to the former USSR and Eastern Europe;
- industrial co-operation agreements with the former USSR and Eastern Europe;
- large scale industrial co-operation agreements with Eastern Europe;
- purchases of licences from the former USSR and Eastern Europe;
- technical co-operation agreements with the former USSR and Eastern Europe;
- the activities of British companies which were either wholly owned or partly owned by commercial organizations based in the former socialist countries of Eastern Europe.

These case studies are then followed by comments, conclusions and suggestions for further research.

Each of the case study descriptions is divided into two sections, in order to obtain an element of longitudinality. The first part summarizes the company's previous trading activities in the Soviet and East European region during the 1960s and 1970s, whilst the second part describes its trading activities in that region during the 1980s and early 1990s.

Notes

1. These previous studies have been published in Hill, M.R. *Export Marketing of Capital Goods to the Socialist Countries of Eastern Europe*, Wilton Publications, Farnborough, 1978; Hill, M.R. *East-West Trade, Industrial Co-operation and Technology Transfer*, Gower, Aldershot, 1983; Hill, M.R. 'Soviet and East European company activity in the United Kingdom and Ireland' in Hamilton, G. (ed.), *Red Multinationals or Red Herrings*? Frances Pinter, London, 1986, pp. 17-87.
2. See Hewett, E.A., *Reforming the Soviet Economy*, The Brookings Institution, Washington, DC., 1988, for a survey of the Soviet economy prior to Gorbachev's appointment as general secretary in 1985, and the reforms introduced in 1985 and 1986.
3. An excellent account of the background to the political and economic changes taking place in the former USSR in the mid to late 1980s can be found in Sakwa, R., *Gorbachev and his Reforms, 1985-1990*, Philip Allan, London, 1990.
4. Hewett (1988), pp. 322-33.
5. *Ibid*, pp. 342-3.
6. See Rollo, J.M.C., 'Western policy: the room for manoeuvre' in Rollo, J.M.C. et al; *The New Eastern Europe : Western Responses*, Pinter, 1990, pp. 100-15.
7. See Batt, J., 'The Political Context', Rollo, J.M.C., Granville, B., 'The economic problem', Rollo, J.M.C., Granville, B., 'The agenda for reform', Malcolm, N., 'How does the Soviet Union fit in', in Rollo, J.M.C. et al (1990), pp. 40-99.
8. Clague, C., Rausser, G.C., (eds.), *The Emergence of Market Economies in Eastern Europe*, Blackwell, Oxford 1992, especially S. Thomas 'The Political Economy of Privatization : Poland, Hungary and Czechoslovakia', pp. 279-95.
9. See Goldman, M.I., *What Went Wrong with Perestroika*, W.W. Norton, London & New York; 1992; Ellman, M., Kontorovich, V., *The Disintegration of the Soviet Economic System*, Routledge, London, 1992.

2 Exports of capital goods

Introduction

The research described in this chapter has been concerned with the activities of a sample of two companies previously involved in the export marketing of capital goods to the USSR and Eastern Europe in the 1960s and 1970s, and a sample of nine companies previously engaged in the transfer of machine tool technology to the USSR over that same time period. The first sample of companies comprised a manufacturer of electrical welding equipment, and a producer of exhaust silencers, drum making machinery and conveyor systems.

A more detailed account of the export trading activities of these companies in the USSR and Eastern Europe is provided in a previous publication of one of the present authors[1] and summarized for the electrical welding equipment company in the first part of that following case study (case study no.1). The exhaust silencer manufacturer has not been described in detail as this company has experienced a dramatic reduction in business from the former Soviet Union since the late 1980s, which compares extremely unfavourably with the early 1970s when exports to the USSR and Poland accounted for some 5-15 per cent of the company's total sales turnover, or some 10-30 per cent of exports. The company has gained some substantial business in Eastern Europe from an agent based in West Germany, but this business has been insufficient to make up for the major shortfall caused by the decline in Soviet orders.

The second sample of companies, engaged in machine tool technology transfer, comprised the following:

- a manufacturer of grinding machines;
- two designers and producers of special purpose transfer machines and link lines for automotive components;
- a designer and builder of special purpose gearcutting machinery;
- a designer and producer of numerically controlled machine tools;
- two designers and producers of automatic turning machines;

- a designer and producer of precision grinding machines;
- a designer and producer of automotive components and of associated production equipment.

A more detailed account of the previous technology transfer activities of these companies in the USSR and Eastern Europe will be found in Hill[2] with a summary of those activities being provided in the first part of each of the following case studies.

Before describing any of these cases in detail, however, it is important to note that the trading activities of five of the companies have changed radically. Two designers and manufacturers of special purpose machining systems had signed a series of contracts during the late 1960s and 1970s for the supply of specialized machine tools, transfer lines and conveyor-linked lines of specially tooled machinery, for the large scale repetitive manufacture of automotive components. The first of these companies ceased trading in the 1980s, whilst the parent group of the other company has progressively sold off its machine tool companies since 1980. It would be almost impossible for the separately owned companies which still exist to now design and build machine tool linked lines, as the overall planning and co-ordination of these projects were carried out by a small team in the parent group's head office.

The third of these companies concerns a firm engaged in the sale of numerically controlled machining centres to the USSR in the mid 1970s, and through the early to mid 1980s, culminating in preliminary discussions with a Soviet machine tool factory over the licensed manufacture of the company's products. The company was taken over by another machine tool producer during 1989-90, and subsequently went into liquidation in 1991. The fourth company was a designer and manufacturer of horizontal and vertical turning, boring and facing machines, which during the late 1960s sold thirteen vertical single spindle machines to the USSR specially tooled for repetitive turning and facing operations, together with associated automated workpiece loading and unloading equipment. These machines were purchased for installation in a new passenger car gear box plant, which was to be completely equipped with imported machinery. This company was taken over, however, and subsequently closed down in the mid 1980s. The fifth company was a designer and builder of high precision thread grinding machines used for a wide range of applications, and also of turning and gearcutting machines. The company sold a large number of each type of these machines to the USSR throughout the 1960s and 1970s, although the company has conducted virtually no export trade in that region since the beginning of 1991.

The latter half of this chapter will therefore provide case study accounts of the remaining four companies which still continue to trade and maintain some interest in exporting to the former USSR and Eastern Europe.

Case study no. 1 - a manufacturer of electrical welding equipment

Commercial relationships with the USSR and Eastern Europe in the 1960s and 1970s

The company described in this case study is a designer and manufacturer of electrical welding equipment which continues to export some 50-60 per cent of its products. In the late 1960s and early 1970s, the company's exports to the socialist countries of Eastern Europe amounted to some 7 per cent of total turnover or 15 per cent of total exports. The majority of the company's exports to the Eastern bloc were to the USSR and included:

- small orders to Mashinoimport[3] from the late 1950s to the late 1960s;
- an order from Mashinoimport in the 1960s for the purchase of special purpose flash welding machines for installation in two car plants;
- a large order from Avtopromimport[4] in the early 1970s for a series of linked machines for the production of arc-welded brake compressor cylinders, and a set of four fully automatic arc welding machines for the production of exhaust manifolds;
- an order for brake-shoe welding machines from Mashinoimport for installation in a car plant in the mid 1970s.

The company also exported specialized welding and heating machines to Poland and gained orders from Hungary and Romania.

Commercial relationships with the former USSR and Eastern Europe in the 1980s and 1990s

The company's order, received in the mid 1970s for an automatic brake shoe welding machine, was the first of seven such machines it exported to Mashinoimport for installation in the Soviet car plant. It continued to sell machines into the Soviet market through the 1980s culminating in a £2.5 million contract for brake shoe welding machinery to Mashinoimport at the end of the decade. Although this contract took many years to negotiate, it provided the UK company with one of its largest single orders for some time. The company felt that it gained the contract because of a lack of local expertise in this specialized area. It was paid in full for this contract through a confirmed letter of credit.

The company's exports to the former USSR have, however, dwindled to virtually nothing over the last two to three years, primarily as a result of Soviet hard currency restrictions. This factor, coupled with the reduced role or complete demise of foreign trade organizations in the former USSR, has forced the company to adapt its trading operations in that market. As Mashinoimport no longer exists, the company has to deal directly with manufacturing enterprises, but until the creditworthiness of these organizations has been established, the company expects to conduct trade indirectly with the member states of the CIS. For example, in 1990, the company supplied eleven small welding machines to another UK company which was exporting its bed-making equipment to the former USSR. The company received payment for its eleven machines despite the fact that the other UK company had apparently failed to receive payment from its Soviet partner, at the time of carrying out this research.

The company gained most of its business with Eastern Europe through the 1980s from Poland and Czechoslovakia, but this has reduced dramatically since the late 1980s and direct exports to the region are now negligible. The company has obtained orders from Poland for £50,000 worth of spare parts and has received numerous enquiries for its equipment, but these orders have not been transformed into actual orders since 1988-89. Business in Czechoslovakia through the 1980s was aided by the activities of the company's agent who had built up a wide range of influential contacts; but the company completed its last contract for Czechoslovakia in 1989-90 - a large order for £350,000 worth of flash butt welding equipment. Despite contractual difficulties arising because the order was initiated before the fall of the communist government and completed afterwards, the equipment was successfully installed and is now in operation. Enquiries for the company's equipment are still being received from Czechoslovakia, but as yet no contracts have been concluded.

The company's commercial ties with Hungary were never strong, but recently it supplied the Hungarian market indirectly with its products. It delivered its drum welding equipment to a Belgian company involved in drum forming machinery, who then supplied to a Hungarian company. The company is currently in touch with other drum forming companies in order to elicit further work of this nature.

The company is confident that business with the former USSR and Eastern Europe will increase in the future and is monitoring developments in the automotive industries of these countries to assess the likely demand for the company's welding equipment.

Case study no. 2 - a manufacturer of grinding machines

Commercial relationships with the USSR and Eastern Europe in the 1960s and 1970s

The company described in this case study is a medium sized engineering manufacturer engaged in the design and production of grinding and honing machinery. It gained substantial business from the Soviet market in the mid to late 1960s: a contract for 300-400 grinding machines delivered between 1967 and 1970 accounted for some 30 per cent of the company's total export turnover during those years. Following the completion of this contract, discussions were initiated in 1971 for the delivery of an additional batch of grinding machines and some honing machines which resulted in another sizeable contract in 1972. A further contract was also awarded to the company in 1974.

Commercial relationships with the former USSR and Eastern Europe in the 1980s and 1990s

The company has experienced a considerable reduction in the amount of business it has gained from the former USSR in recent years, with no direct delivery being undertaken since 1987, although the company is still indirectly supplying the former USSR but in much smaller quantities. In August 1991, however, a contract was signed with a ball-bearing factory for the supply of equipment worth some £0.75m, but the fulfilment of this contract had not been commenced at the time of carrying out this research as the customer had insufficient hard currency reserves at that time, although the equipment was needed. The contract has been referred back to the Economic Planning Committee of Ukraine and the company is hoping for eventual ratification.

A similar situation has evolved in Eastern Europe over the last few years, with the company receiving few enquiries or orders. It was felt that the main reason for the dramatic decline in the company's business in the market was the lack of hard currency available from prospective customers rather than any reduction in the demand for the company's products.

Case study no. 3 - a producer of automatic turning machines

Commercial relationships with the USSR and Eastern Europe in the 1960s and 1970s

This company is a designer and producer of multi-spindle automatic turning machines. It delivered a total of eighty of these machines, with associated specialized tooling, to a Soviet foreign trade organization in the late 1960s and early 1970s. These machines were purchased for installation in a new Soviet passenger car factory as part of the Soviet government's decision to rapidly expand its passenger car production at that time. It was decided that design and production know-how would have to be purchased from a Western company if the USSR was to attempt to meet its production targets for contemporary models of passenger cars. This was due to the relative backwardness of Soviet automotive design and production techniques in the late 1960s.

Commercial relationships with the former USSR and Eastern Europe in the 1980s and 1990s

This company has since expanded its product range in the last two decades, and it now manufactures manually operated optical profile grinding machines, CNC optical profile grinding machines and vertical turning machines as well as its multi-spindle automatic turning machines.

Although the company has increased its product range, it has dramatically reduced its workforce from around 2,000-2,500 employees at the peak of its operations in the 1970s, to around 150 people employed currently.[5] Furthermore, the company now exports approximately 90 per cent of its products, whereas exports previously amounted to around 30 per cent of overall business. These changes in the company's mode of operation were attributed to increased automation which has reduced manpower requirements; and to the gradual demise of the British engineering industry, evident since the 1970s, which has made the company almost entirely dependent on export markets.

The company has failed to gain any further business from the Soviet automobile factory since its last order in the 1960s, and has not obtained a significant amount of business in either the former USSR or Eastern Europe during the last two decades. The occasional individual sale of its machines has materialized, however, as a result of the work of an agent working on its behalf in Poland and the company regularly quotes for this agent, but its prices are generally too expensive for Eastern European buyers. Talks were also held with a Soviet delegation who visited the company in 1990, to discuss the possible purchase of the company's technology. The delegation, which included the production director and the chief engineer of a large Russian factory and a senior bureaucrat, admitted that it had no access to hard currency and hoped to pay for the company's designs, tooling and know-how in Soviet machines. The UK company was prepared to consider concluding a deal of this nature, but was never contacted again by this delegation. The largest and most recent contract secured by the company from this region, was an order to supply thirteen multi-spindle automatic lathes of differing specifications to East Germany. This contract was awarded by the East German authorities in 1990, over more expensive bids submitted by some of its competitors.

The company felt that this contract had been both complicated and time consuming to undertake. Difficulties had emerged because of the lack of certain technical developments in East German tooling, which meant that the company was unable to supply the carbide tooling which predominates in the West. Instead, the company had to provide old fashioned high speed steel tooling which could be more readily replaced in East Germany. The company also had to specify alternative cutting oils as its preferred cutting fluids were not manufactured in East Germany and therefore could not be replenished domestically by the customer. Furthermore, the machineability of the East German raw materials was of a very poor standard which made them difficult to work with effectively. Thus, the company had to deliver thirteen machines of varying specifications, without the benefits of its standard tooling, lubrication or raw materials.

Despite these complications, the company did successfully complete its role in the contract and the machines were all installed in six different factories in East Germany. Full payment in hard currency was received from the East German authorities and the company also agreed to counterpurchase two East German machines from a UK distributor. The purchases of these machines, which were cheaper and less technologically advanced than their Western counterparts, amounted to around 5 per cent of the total deliveries from the UK firm. The company does not expect any follow on business to materialize from this source, however, because another similar contract would now be most likely undertaken by companies in the former West German state, following unification.

Case study no. 4 - a producer of gear cutting machines

Commercial relationships with the USSR and Eastern Europe in the 1960s and 1970s

This company is a designer and manufacturer of gearcutting machine tools and gearcutting tool grinding machines. The company sold a large quantity of its standard machinery (ie. gear hobbing machines and single-spindle shaping machines) to the USSR in the 1960s and 1970s; and has also sold twin-spindle gear shaping machines, specially tooled and equipped with component transfer equipment, into the same market. It sold thirty-six of this latter type of machine to the USSR in the 1970s. The machinery was specially tooled and equipped for cutting a series of gear-teeth on a truck layshaft at the newest and largest Soviet truck factory of the time (Kamaz).

Commercial relationships with the former USSR and Eastern Europe in the 1980s and 1990s

The company has experienced significant changes since the mid-late 1970s in terms of its trade with the former socialist countries of Eastern Europe, and also in its own internal structure. Twenty years ago, the USSR represented one of this company's main export markets amounting to around 20 per cent of its business, but now the company's trade with the Commonwealth of Independent States has declined to virtually nothing. At the same time the company has been affected by the gradual demise of mechanical engineering in the UK, and the workforce has been dramatically reduced from 200 in the late 1970s to eighteen today, although the company's products are similar.

The Soviet purchases of the company's machines continued through the 1980s, albeit on a far smaller scale than previously. Only two or three large, five foot diameter, herringbone gearcutting machines were sold into that market throughout the decade. The company gained these contracts through Stankoimport[6] by virtue of its reputation as a key supplier to the Soviet market, and because the buyer would have found it difficult to find another manufacturer producing these types of machines. However, the company sold its last machine to the USSR in 1987-8. The significant reduction in the company's sales to this market was blamed primarily on the decentralization which was commenced in the USSR in 1987, rather than on any decline in demand for the company's products in that country. Financial obstacles have curtailed and prevented trade as individual Soviet enterprises have seldom had the funds available to take advantage of their new autonomy. The situation, it was felt, was not aided by the reluctance of banks and financial institutions to provide the necessary financial backing to help projects get off the ground. Although the company is confident that it will regain business eventually, recent discussions with Soviet manufacturers like Lada have failed to yield any concrete orders.

Throughout the 1980s, the company also gained business in Eastern Europe. During the early 1980s a variety of the company's machines were sold to the Ursus tractor factory in Poland, the last orders being delivered in 1984. In 1987, the company was awarded a contract worth £3-4 million overall, to retool the Wartburg car factory in East Germany by supplying fourteen machines. However, since unification, many of the old East German factories have closed as they were inefficient and unprofitable when operating under market conditions in the new unified German state. The aforementioned car factory was one of the plants to be so affected and consequently all the company's previously delivered machines are currently available for sale in the second hand market.

In the past two to three years the company has dramatically reduced both the amount of time its personnel spend in the former socialist countries of Eastern Europe, and its marketing activity in that region, reflecting its belief in the reduced opportunities for business in that area for the foreseeable future. The company also feels sufficiently confident that its reputation as a reliable supplier to the market in the past would ensure that it would be contacted should an organization in either the former USSR or Eastern Europe require any of its machines. This was illustrated recently when, in 1991, a substantial contract was gained from Romania, a country to which the company has sold many of its machines since the 1960s. This agreement involved the supply of £150,000 - £200,000 worth of the company's cutting tools suitable for use in the production of oil equipment. The company was awarded the contract directly from the end user, Vulkan, a manufacturer of pumping units for the oil industry. As Vulkan exports many of its end products it has a supply of hard currency with which to purchase Western machinery. The contract was organized through the British based Anglo-Romanian Bank and payment was received in full by the company through an irrevocable letter of credit. The machinery is being operated effectively in Romania and the company felt that a good working relationship had been established between the British and Romanian personnel involved in the deal, heightening the probability of full payment for this agreement, and also enhancing the prospect of further business for the British side from this Romanian enterprise.

Case study no. 5 - a producer of automotive components

Commercial relationships with the USSR in the 1960s and 1970s

This company is a designer and manufacturer of automotive components. During the late 1960s the USSR purchased a licence for the production of a clutch unit of the company's design, together with production know-how including some plant and tooling. The Soviet manufacturers using these clutches included Avto-Vaz, the manufacturer of Lada cars. The contract expired in 1988-89 having run its course, and by that time the product in question was being manufactured for use in the Lada Samara car. The main aim of the Soviet side in signing the original contract was therefore achieved.

Commercial relationships with the former USSR in the 1980s and 1990s

As a direct result of the successful achievement of the previous contract, the company was asked by the same ministry and foreign trade organization to submit proposals for another similar project involving different end users - Kamaz and Gaz, manufacturers of trucks in the Russian republic. A contract was signed in 1984 which required the company to:

- design a range of clutches for a wide variety of different commercial vehicles;
- manufacture and provide prototypes of the clutches to be tested by the Soviet manufacturers;
- supply the manufacturing technology used to complete the prototypes, including the component and tooling drawings, plant layouts and material specifications;
- specify the production equipment the Soviet side should acquire in order to manufacture the products and recommend the sources of supply.

Although the company has completed the work on the above four areas and provided the Soviet side with the necessary documentation, Kamaz has yet to achieve production because of its inability to purchase all of the recommended equipment.

The financial package worked out for this deal involved regular payments to be made by the Soviet side to cover the cost of the licence fee and the work undertaken by the company in product design and preparation of manufacturing documentation. Initially, payments were made in full at appropriate intervals, but, in the two years prior to carrying out this research, the situation had deteriorated with no payments being received in 1991. The company is maintaining regular contact with the foreign trade organization and end users to monitor developments and provide advice.

The company is not too discouraged by the current status of this agreement and still sees further business opportunities in the future. Within the last two years it has submitted three new proposals involving the transfer of the company's technology to organizations in the former Soviet Union. One of these proposals was turned down, while the other two remain pending due to the political and economic changes in the country. The company feels that the main change it has experienced in doing business with the former USSR is that whereas previously emphasis was placed on the technical content of the proposal, the emphasis has now moved to its commercial aspects, including external financing.

Commercial relationships with Eastern Europe since the 1960s

There was very little business conducted in Eastern Europe in the 1960s and 1970s because of the difficulty of selling finished components to that market at that time. As these countries have become liberalized and able to trade openly with the West, they have increased their purchase of vehicles, both new and used, outside the former Comecon bloc. This has created a market opportunity for the company as local manufacturers are often unable to supply the required spare parts. Accordingly, the company has made some progress in selling its products in Poland and Hungary and by appointing agents in both countries with the responsibility to achieve further sales of this nature.

The company, however, sees the most likely long-term business in these countries as being licensing or direct investment and has therefore initiated contacts with a number of companies in Poland, Hungary and Czechoslovakia, the three countries it considers to offer the most favourable prospects for future trade. To date no contracts have been concluded.

Comments

The information obtained from the case studies of the companies in this sample suggests a sharp decline in the exports of capital equipment and machine tool technology by British companies to the former socialist countries of Eastern Europe, compared with the levels of export deliveries some ten or fifteen years previously. To a great extent this has been caused by major reductions in demand from Soviet customers, who had previously been important buyers. The major reason advanced for these changes in the level of demand was an acute shortage of hard currency by Soviet buyers, with this shortage more apparent in the individual factories now responsible for their own foreign trading activity. These decreases in demand, combined with concerns over Soviet factory creditworthiness have led to some companies re-organizing their export policies towards selling through Western third parties rather than direct to the customer.

It is apparent, however, that these conclusions have been reached from the study of an extremely small sample, selling to a narrow customer profile as many of the companies sold their products to the Soviet automotive industry. This sector experienced an investment boom in the late 1960s and 1970s, but there has been a reduced level of follow up investment since that time. The one exception to this possible criticism of narrow customer profile was the grinding machine company which sold its products to a wide range of engineering customers, although the ball-bearing industry was clearly an important buyer, and major customers for that industry would also be automotive and agricultural engineering. In view of the smallness of the sample and the purchasing behaviour of the customer industries, the studies reported in the subsequent chapters of this book are therefore particularly relevant from the viewpoint of confirmation or rejection of these initial findings.

One of the major points emerging from this research, however, has been the rapid reduction of the number of factories within the British machine tool industry, and the reduced numbers of employees in those factories that remain. In this particular study, for example, the number of factories had halved, whilst the number of employees in some remaining factories had been reduced by a factor of ten. These reductions were caused partly by the severe recession in the British mechanical engineering industry in particular and the world recession in general, causing many user factories to re-equip with second hand machine tools obtained from those factories which had been forced to close.

It remains to be seen, therefore, whether the British machine tool industry will be able to reap any benefits from any upturns in future orders which may occur in the former USSR and Eastern Europe. As many factories have been closed, and skilled workers lost, it is unlikely that any of these factories can be rapidly re-opened because of the lead times required to recruit labour and install machinery. Furthermore, it remains to be seen whether those factories remaining will be able to benefit substantially from any major upturn in business, as the international market continues to be extremely competitive. Those companies that do remain in this sample, however, can offer expertise gained from the previous successful completion of large contracts in the former Soviet Union, together with the capability to provide contemporary technology. One company has also been engaged in a technology transfer arrangement with a large Soviet truck factory, which is likely to have some access to hard currency, as explained in case study no.4 in the following chapter.

Notes

1. Hill, M.R., *Export Marketing of Capital Goods to the Socialist Countries of Eastern Europe*, Wilton Publications, Farnborough, 1978, pp. 92-6, 114-6.
2. *Ibid*, pp. 85-91 and Hill, M.R., *East-West Trade, Industrial Co-operation and Technology Transfer*, Gower, Aldershot, 1983, pp. 49-74.
3. The Soviet foreign trade organization specializing in the import of various types of industrial machinery, including welding equipment.
4. The Soviet foreign trade organization specializing in the import of equipment for the motor industry, particularly for large new factories.
5. Since completing this research and writing this case, the authors have seen reports that this company has been sold. The automatic turning machine business is reported to have been purchased by a company engaged in the rebuilding of these machines, employing nineteen personnel.
6. The Soviet foreign trade organization specializing in the import of machine tools.

3 Industrial co-operation

Introduction

This chapter describes the activities of a sample of fourteen companies which have been partners in industrial co-operation arrangements with the USSR and Eastern Europe, with ten companies described in some detail as they have continued their operation in this market since the previous research some ten years ago. For the purposes of this research, industrial co-operation was viewed as a set of trading relationships, which extended over a number of years and which went beyond the straightforward sale or purchase of goods and services, to include a set of complementary operations. These could be in the areas of production, product development, transfer of technology, or marketing. In practice, therefore, an industrial co-operation agreement usually included the following elements:

- a transfer of technology from the British partner to the Eastern partner for the design, development and/or manufacture of a particular finished product. The technology transfer may also have included the sale of appropriate licences, manufacturing equipment or components, and the training of personnel;
- the time interval of the industrial co-operation agreement was usually longer than that associated with a one-off sale;
- if appropriate, payment for the technology transferred was made partly in items related to the finished product.

The sample of companies chosen was diverse, and all had been engaged in the exporting of capital goods to the USSR and Eastern Europe during the 1960s and 1970s. Three of the companies also signed industrial co-operation agreements with the USSR towards the end of the 1970s, whilst another ten companies had signed similar agreements with Eastern Europe over the same time interval. In one case, however, the company did not become involved in industrial co-operation activities until the late 1980s and early 1990s.

The sample of companies comprised the following:

- a toy company;
- an automotive and aerospace engineering company;
- a designer and manufacturer of automotive shock absorbers;
- a designer and manufacturer of power transformers;
- a manufacturer of photocopying equipment;
- a multinational trader in semi-manufactures, machinery and raw materials;
- an international group of engineering companies;
- a producer of cranes;
- a designer and manufacturer of mining equipment;
- a manufacturer of transport equipment and associated components;
- a manufacturer of computer equipment;
- a nuclear engineering company;
- a marketing consultancy operating primarily in the agricultural sector;
- an aero-engine manufacturer.

A detailed account of the companies' export trading and industrial co-operation activities in the 1960s and 1970s can be found in previous publications by one of the present authors.[1,2] A summary of those activities is provided in the first part of each of the following case studies, followed by an account of their activities in the 1980s and 1990s in the second part of each case study.

In the case of some companies, it was found during the follow up research that their business relationships had ceased in the USSR and Eastern Europe, usually because of a change in company ownership, or because they had ceased to trade altogether. Such cases included a subsidiary of a major toy group in the UK which was set up specifically to handle an industrial co-operation deal with the USSR, signed in 1975. The arrangement consisted of a ten year compensation deal to supply the Soviet Union with tooling, ancillary equipment and materials, with the ultimate objective of establishing a toy manufacturing capacity based on Western designs. In 1981, however, the American-based parent company went into liquidation, and the receivers closed down all of its UK subsidiaries. The Soviet side paid off all its financial obligations in full, thus concluding the agreement between the two sides.

A second case was that of a large company which had manufacturing, marketing and service facilities both in the UK and overseas. Each operating company within the divisionalized overall structure carried out its own marketing, and establishment of trade and co-operation agreements, usually having one manager responsible for Eastern Europe. The company also had a central division responsible for co-ordinating trading activities with the Soviet Union. The company held discussions on technology transfer and industrial co-operation in aero-engine component technology with a prospective Soviet buyer during the late 1970s, but the agreement was never concluded because of concerns by the British government that the technology transferred may have had a military application. The company, did, however enter into an industrial co-operation agreement with a Polish foreign trade organization in 1978. This agreement included the sale of a licence and the provision of relevant design and technical information for a tractor alternator, with possible subsequent variants for use in cars and light aircraft. The licence agreement was signed in the late 1980s, and no other new contracts were signed. The company had also divided into more divisions at the time of this present research but no evidence could be found of any central co-ordination of the East European market, possibly because the amount of trade conducted by each division in this market has reduced considerably.

A third case concerned a company which had signed two industrial co-operation agreements for telescopic shock absorbers with Romanian and Polish foreign trade organizations in 1966 and 1968 respectively. Both agreements were for a period of ten years, and have subsequently expired. The company was taken over in 1989 by a Belgian company, and responsibility for trading relationships with the USSR and Eastern Europe has now been transferred to the company's head office in Brussels.

A fourth, and final case, concerned a company that had sold its products to Poland and Czechoslovakia and signed an industrial co-operation agreement with a Hungarian foreign trade organization responsible for the import and export of electrical transmission equipment. The agreement, which related to the provision of technical know-how in the design and manufacture of power transformers, was originally signed in 1967 and subsequently extended for another five years until 1977. The agreement involved the sale of a licence, and the provision of associated design documentation and manufacturing advice. Although the British company continues to sell technological know-how to the Hungarian organization, the company has not developed any other trade links with Soviet or East European organizations.

The remainder of this chapter describes those companies which were previously studied some ten to fifteen years ago, and still continue to trade actively with the former USSR and Eastern Europe.

Case study no. 1 - a manufacturer of photocopying equipment

Commercial relationships with the USSR and Eastern Europe in the 1960s and 1970s

This company's main business involves the design, manufacture and sale of photocopying equipment, but its product range also includes typewriters and printers. The company has had extensive experience in selling to the former socialist countries of Eastern Europe and has consistently devoted considerable resources to the development of these markets.

Throughout the 1960s and 1970s, the market for photocopying equipment in the USSR and Eastern Europe was severely restricted because of governmental fears that photocopying machines could be used as underground printing presses. However, the company placed great importance on developing this market and it reportedly held some 85 per cent of the Comecon copier market during the 1970s. It carried out extensive market research in each of the then socialist countries, considering each country as a separate unit with its own buying policy. The main sources of market information were considered to be the company's local agents and local representatives situated in each country, although sales were made direct either to a foreign trade organization or to an end user who was legally entitled to purchase. The company increased the resources it committed to the market over this period, with the number of personnel employed in the East European Operations Office expanding from two in the 1960s, to 115 by the mid 1970s. Furthermore, the company avidly promoted its products through seminars designed to illustrate the variety of applications of the company's products, and exhibitions to demonstrate the operation of the machines. These promotional activities were supplemented by the activities of the 'exhibition train' and 'exhibition trailer' mobile exhibitions, which contained selected products and information on the company's product range and which travelled to various end-user sites to demonstrate and display the company's products.

Commercial relationships with the former USSR and Eastern Europe in the 1980s and 1990s

The company's sales to the former socialist countries of Eastern Europe amounted to £50 million in 1989. This represents a similar proportion of total annual sales turnover to that of twenty years ago - around 2 per cent. The company has lost its market share in these countries, however, and considers that it now has roughly 40 per cent of the market in the former USSR, 60 per cent in Romania and Bulgaria, and around 10 per cent in Poland, Czechoslovakia and Hungary. This reduction in market share is attributed to the influx of Japanese competitors into the market, and in particular to their introduction of the cheaper personal photocopier.

However, the company is optimistic about its future sales to the CIS and Eastern Europe because the liberalization of this market in the mid 1980s ended the prevailing restrictions on the purchase of photocopying equipment. Furthermore, these restrictions left the former Eastern bloc with a dearth of photocopying equipment, of which the company hopes to take advantage. Currently there are 3,000 copiers for every million inhabitants in Western Europe, 300 for every million inhabitants in Latin America and only 40 for every million inhabitants in Eastern Europe and the CIS. The extent of this demand has encouraged the company to allocate more time and resources to this market-place and current company policy is to effect a four fold increase in sales by 1995.

The company felt that the other main effect of the liberalization of the market was the reduction in the role of the foreign trade organization as the customer. In every state with the exception of Russia, these organizations have either collapsed or dramatically changed shape and individual enterprises have been given more autonomy to make their own purchases. This has forced the company to adapt its methods of conducting business in Eastern Europe and the CIS, as the percentage of company sales to that market conducted through foreign trade organizations reduced dramatically from 38 per cent to 2 per cent during 1991.

However, although the demand for copying equipment in this market is insatiable, the inaccessibility of hard currency is restricting its widespread purchase. The company, therefore, has allocated considerable resources to identify potential customers and attract their future business. For example, in every country in the former Eastern bloc either a new office was established (as with those in Warsaw, Prague, Budapest, Kiev and Riga) or the existing office was expanded (as with Moscow, Bucharest, Sofia and Belgrade). These offices are run by teams of handpicked local and non-local personnel who are all linguistically capable, entrepreneurially-minded and have some understanding of market economics.

The company is also developing a new distribution system and is recruiting new dealers to promote and sell the company's products to customers. The drive to recruit new dealers was launched in November 1991 as part of the company's promotional 'travelling roadshow' which visited twelve cities throughout the region, including Yugoslavia. So far, ninety of the targeted number of 170 dealers have been recruited, and, in 1991 the proportion of total business in the region gained through the dealers rose from zero to 38 per cent during the course of the year.

The company found it relatively difficult, however, to recruit personnel with real commercial awareness or knowledge of Western sales techniques. Hence, the company has initiated its own training scheme through two new joint ventures - one in Brno, Czechoslovakia and one in Moscow - to instruct its staff in Western business practices. The Czechoslovak company provides a training course covering sales and service techniques and technical instruction on the maintenance requirements of the machines, which all the company's dealers and employees must attend. The Moscow joint venture was signed with the National Institute for the Graphic Arts, and provides the company with two floors of office space inside the Institute in which to run training courses for selected Russian personnel, the most promising of which will be recruited to the company. In a similar vein, the company has set up business courses in association with Manchester Business School and the British Council and is currently providing tuition for around 25 students from St. Petersburg. Furthermore, a Russian, a Pole and a Hungarian are all working for the year in the financial and administration departments of the company's head office in London.

The company has also expanded its operations through the opening up of new copy shops offering similar services as the initial one set up in Moscow as a joint venture in 1989, which has proved to be a commercial success. The company now fully owns this copy shop having bought out its non-commercial partner when the law permitted it to do so. There are now eight of these copy shops throughout the former Eastern bloc. As well as providing a variety of copying and stationery services to businesses and individuals, these shops also serve to illustrate the capabilities of the company's products and thus encourage future sales.

The company is optimistic about its future prospects in the market-place, particularly in Romania, where the company is experiencing its highest rate of growth in its sales, Czechoslovakia (second highest growth), and Poland (largest absolute levels), but it is also positive about the longer term prospects in the former USSR, citing the Baltic and the Southern Asiatic states as those offering the most potential for business.

Case study no. 2 - a multinational trader in semi-manufactures, machinery and raw materials

Commercial relationships with the USSR and Eastern Europe in the 1960s and 1970s

This is a division of a company which is a long established multinational trader in semi-manufactures, machinery and raw materials. It has engaged in trade with both Russia and the USSR for over 80 years and has maintained a permanent presence in Moscow since 1962, engaging in the trade of raw materials and chemicals. In addition to bilateral sales, the company acts as agents for a large number of principals, receiving commission from them according to volume of sales.

In the late 1970s the company negotiated a technical and marketing agreement with the 'Tekhmashexport' Soviet foreign trade organization, and the Soviet Ministry of the Machine Tool Industry. The subject of the agreement was a Soviet produced injection moulding machine which incorporated the British company's control and design features. This agreement, which was for five years with renewal if necessary, involved:

- training of Soviet personnel;
- supply of parts to the USSR;
- provision by the Soviet partner of products produced to the company's, specifications to be marketed in the UK by the company;
- an agreement for marketing and servicing in specified geographic areas;
- co-operation in research and development;
- exchange of scientific and technical information.

The company entered this agreement to achieve the expansion and diversification of future market opportunities. By involving itself in an industrial co-operation agreement principally designed to secure foreign currency earnings for the USSR, the company expected to further a business relationship whereby future sales of other company products to the USSR would be greatly facilitated.

Commercial relationships with the former USSR in the 1980s and 1990s

The above-cited technical and marketing agreement ended in the mid-1980s, leading to a culmination of the company's involvement in Soviet machine tools. The company had become agents for the Soviet machine tool industry only as a temporary measure, and, once its obligation to the Soviet side was completed, it returned to operating in more familiar territory.

The company has experienced both a change in direction, and a reduction in the level of its trade with the former USSR, as a result of the political and economic upheaval prevailing in that country since the late 1980s. The devolution of foreign trade competences to manufacturing enterprises has created problems for the company because an enterprise generally has limited hard currency resources and its personnel scant experience in conducting foreign trade. Under the previous system of centralized operation of foreign trade, the state buying organizations would issue an enquiry and negotiate a contract; and the company would be assured of payment for the goods in hard currency, for successful completion of the project. Now, however, the company has to develop alternative arrangements to overcome the scarcity of hard currency, usually through a barter arrangement which can be difficult and time-consuming to organize. Although the company's dealings with the USSR in the past were characterized by over-bureaucratization, payment in hard currency was always achieved, which cannot be guaranteed in today's decentralized and rapidly changing conditions.

Furthermore, the devolution of power to the factory level and the consequent reduced role of central organizations in the conduct of foreign trade has meant that many of the contacts the company had established in the appropriate ministries over the years no longer wield the same influence today. In seeking to establish new contacts, the company's efforts have been hindered by the inadequate communications infrastructure currently existing in the former USSR. To a company also experienced in trading with the former USSR as a monolithic entity, the current fragmentation of the union has also presented problems. The existence of different rules and regulations in each republic has meant that an overall company strategy towards the region is not applicable today.

All the above factors have resulted in a contraction in the company's business transactions with the former USSR. The company felt that while the changes in the former USSR would be good for trade eventually, it would be some time before beneficial effects were felt by it and other UK companies.

Case study no. 3 - an international group of engineering companies

Introduction

This case study describes an international group of engineering companies which has had commercial links with Russia dating back to 1840. The companies in the group with experience in the Soviet and East European market are Engineering and Construction, Power Engineering, Offshore Construction and Plastics Machinery. Since 1945 these companies have completed a wide variety of projects in the USSR involving engineering, equipment supply, the construction of chemical plants, plastic coating and film extrusion lines, machine tools and the supply of a total of sixty-eight large industrial gas turbines. In recent years this company has also supplied the USSR with complete automated assembly systems for a variety of products, mostly for the automotive and agricultural machinery industries. Many of these contracts involved counter-trade arrangements under which this company and its Western partners purchased substantial quantities of chemicals and other products from the USSR.

The value of Soviet business to the group has been substantial and, since 1974, it has recently won new contracts in the USSR worth approximately £450 million. This business has been distributed throughout the group's companies as described in the following sub-sections.

1. Engineering and construction

Commercial relationships with the USSR in the 1960s and 1970s This international company operates primarily in the oil, gas, polymer, petrochemical, pharmaceutical and environmental protection industries providing engineering, procurement and construction management services. Since 1960 this company has been active in the USSR through design and supervision of the construction of chemical plants. During these two decades the company oversaw the construction of fourteen chemical plants in the USSR with contract values for each project ranging from £0.32 million to £57.74 million This provided very satisfactory and profitable business over the years, the only complaint on the British side being the length of time taken to build the plants - on average seven to eight years. Some of the contracts involved a separate buy-back commitment which usually made up around 10 per cent of the total value of the contract, but this was never a pre-condition for the deal progressing. The company would receive payment from the central budget derived from the state sale of oil, gas and petrochemicals.

Commercial relationships with the former USSR in the 1980s and 1990s The nature of this business was affected with the election of Gorbachev to the post of General Secretary of the Communist Party of the Soviet Union in 1985. One of the main effects, noticed by all companies trading in the USSR, was the considerable pressure from the Soviet government for them to enter into joint ventures with Soviet organizations. The main attraction of joint ventures to the USSR was that the foreign partner would provide the necessary finance to ensure the project could be implemented outside the confines of the current Soviet hard currency shortage. Previously, the foreign trade organizations of the Ministry of Foreign Trade dominated and controlled the whole foreign trade process, with end users kept at arms length. Along with other contractors, this company was asked to make proposals for the supply, construction and subsequent management of complete process plants, which would involve the marketing of the final product outside the USSR. This proposition was not completely relevant to the company who are not in the business of operating plants and selling products. However, in the final stages of the successful negotiation to work on its sixteenth chemical plant in the USSR in 1987 - a new 100,000 ton/year polypropylene plant project for Stavropol Polymer (a Soviet enterprise within the Ministry of Chemical Industry) at Budyennovsk, the company did commit itself to collaborate on a particular project. The outcome of these discussions was the formation of an unusual joint venture company, which was the first Soviet-British engineering and trading joint venture.

This joint venture, signed in April 1988, was set up by the UK company together with representatives from three organizations within the Ministry of the Chemical Industry, the Moscow Narodny Bank and a UK bank (Morgan Grenfell). It is now in the process of organizing the modernization and extension of the polyethylene plant at Budyennovsk (to expand ethylene and polyethylene capacity from 200,000 ton/year to 300,000 ton/year). This will be followed by a similar project involving the improvement of productivity at a polyethylene plant at Kazan (in two stages from 200,000 ton/year to 400,000 ton/year). Both of these plants were originally engineered and supplied by the UK company, using the technology of the Union Carbide Corporation of the USA. The value of business for the company resulting from these first two projects will be about £130 million.

The company's joint venture concept involves a circular set of deals which are all dependent on each other. For example, the following process is undertaken in the modernization of Budyennovsk:

- Stavropolpolymer contracts to deliver polyethylene to the joint venture;
- the joint venture engages Union Carbide to sell the polyethylene on the world market;
- the joint venture borrows money from banks (using receipts from future polyethylene sales as security);
- the joint venture uses the money to pay Stavropolpolymer in advance for future polyethylene deliveries;
- Stavropolploymer uses the money to pay the UK company for their services, and Linde to modernize the plant which makes the polyethylene;
- when the plant is completed, Stavropolpolymer starts producing polyethylene and the resulting income is, over a period of nine years, used to repay the loan.

The total loan value, raised from a syndicate of twelve to thirteen banks who all became shareholders in the venture, was $330 million (including facilities which are used only in the event of certain contingencies). The joint venture, and not a former Soviet organization, is therefore the borrower. The advantage for the Russian shareholder is that the project can proceed outside the constraints of central hard currency funding. Work is well underway on this project with progress said to be two to three times faster than on similar Soviet projects. Design is complete, site work is well advanced and the Russian subcontractor, ZSM, is being paid partly in hard currency and is achieving close to Western productivity levels. This compares with the far slower workrate being achieved on a polyproplyene plant at Budyennovsk, which is also being supplied by the UK company and Union Carbide, but remains outside the joint venture arrangements.

The joint venture model is also currently being used in the attempt to construct a greenfield petrochemical complex at Novyi Urengoi, Western Siberia. This joint venture company, was formed by the UK company with Gazprom (the ministry responsible for the gas industry within the Russian Republic) and Linde who are to act as the exclusive sub-contractor for the gas cracker. It is expected that the UK company and Gazprom will each hold a 30 per cent stake in the venture, and a group of banks will share the remaining 40 per cent. The UK company is to manage the project and provide engineering procurement and services for the polyethylene plant using Union Carbide's Unipol process. In addition to supplying technology, Union Carbide, not a joint venture partner, will act as the marketing arm for the joint venture. This agreement is still in its formative stages and although the joint-venture company was set up in June 1990 and the preliminary work has begun, the finance to pay for the major construction has still to be raised.

The company believes that as a means to maintain a strong Russian business interest and ensure foreign currency earnings, the joint venture-type model has, to date, proved to be a positive structure for reacting to the dramatic changes in the former USSR. Thus, it is expected that more deals will be formulated using this joint venture formula in the future.

However, the company felt that its long term future trading ties with the former USSR would also involve the establishment of joint venture companies providing engineering services within that region. Hence, in December 1990, a third joint venture was set up by the UK company with Giproplast (the project engineering design institute for the polymers industry) as the principal Soviet partner. A few other Soviet enterprises were involved, but no banks. This joint venture company provides engineering services specifically for the chemical industry in the USSR, with particular emphasis on the polymer sector. It is benefiting from cheaper labour in the CIS and has already carried out some work in the former USSR, and also aims to sell some products to the West to ensure a hard currency income. In the same vein, the company set up a fourth joint-venture towards the end of 1991, to serve the Soviet pharmaceutical industry and eventually the food industry.

The company felt that the advantage of this type of structure was that it offered the UK partner cheaper manpower costs and the CIS side a more effective transfer of technology. Furthermore, the companies formed are service industries, the setting up of which does not involve the same risks or costs involved as in a manufacturing industry. It was felt that this factor heightened the chances for their survival in the short term and their success in the long term.

2. Power engineering

Commercial relationships with the former USSR in the 1980s and 1990s
This company has supplied turnkey power stations to forty-eight countries and generated export business worth more than £2bn. It manufactures large heavy duty gas turbines and operates as a specialist management contractor for power generation projects based on gas, diesel, steam or hydropower. It has performed very successfully in the Soviet market since 1975, selling sixty eight large industrial gas turbines (to the design of General Electric of the USA) to the USSR - all for pumping gas along Soviet pipelines.

This business was affected by the operation of US export regulations resulting from the imposition of martial law in Poland in 1981,[3] and also by Soviet hard currency shortages in the mid 1980s. In December 1981, the American government reacted to events in Poland by introducing an embargo on equipment for the pipeline from Urengoi to Western Europe.

A ban was placed on the sale from the US of General Electric rotors, stator blades and nozzles for use in turbines to be delivered for installation in that pipeline. Since this company required these components in its products, it was finding it almost impossible to meet its contractual delivery requirements of twenty-one large industrial gas turbines together with associated spares to its Soviet customers in 1982. The situation was exacerbated in June 1982 by further implementation of the US Export Administration Act which expected this company and other European firms in a similar position, to conform to US requirements for embargoes to specified markets, even when the companies had stocks of previously sourced American components and technology. The British government, invoking the British Protection of Trading Interests Act, then ordered the company to disobey the American conditions. This obedience to its own national government laid the company open to possible loss of US markets and legal penalties on its US assets and personnel. The company felt, however, that it was more important to fulfil its contractual obligations to its Soviet customer, a decision which was considered favourably in the USSR.

The last few years has seen a marked reduction in this business because the previous Soviet customers are now purchasing domestically-produced turbines. The prohibitive cost in importing Western turbines and the lack of Soviet hard currency reserves has meant that since 1985-86 the Soviets have been forced to use turbines manufactured in the USSR, although their technical performance may not match that of the company's products.

3. Offshore construction

Commercial relationships with the former USSR in the 1980s and 1990s
This is one of the largest, fully integrated offshore construction contractors in Europe. It is also involved in the UK onshore electrical and mechanical contracting industry.

This company does not have the same long established ties with the former USSR as the previous two in the group, but recently it has begun to work in association with the group's engineering and construction company and fabrication yards in Baku, to make piles for a British oil company's platform operating in the North Sea (see chapter 6, case study no.4). The deal provided the Soviet side with an initial £7 million contract and the potential for future business as the long term aim is to fabricate the complete platform in the former USSR. The Baku yards are very experienced in fabrication and the components would be prepared there, but the final construction of the complete platform would have to be in a Baltic port, perhaps Kaliningrad. This deal would appear to benefit all sides as it provides the CIS side with a rare export opportunity in a manufactured product, Offshore Construction with a trading opportunity in a new market and the British oil company with confidence in its final purchase, resulting from the involvement of a reputable UK company in the deal.

4. Plastics machinery

Commercial relationships with the USSR in the 1960s and 1970s This company provides processing machinery, systems and complete plants for the plastics industry. It has had some experience in the USSR namely supplying, installing and commissioning two kraft paper coating lines to Tekhmashimport in 1959; supplying, installing and commissioning three coating lines to Prommashimport in 1967; supplying extrusion coating equipment via Tetrapak, Sweden to Prommashimport in 1970; supplying, installing and commissioning a polyethylene film coating line to Prommashimport via Sim-Chem for Vladimir in 1971 and supplying eighteen presses for foam mattress manufacture in 1977.

Commercial relationships with the former USSR and Eastern Europe in the 1980s and 1990s In 1988 this company secured a contract, worth around $13m, to supply thirty thermoforming machines for production of pvc margarine tubs and lids. This formed part of a programme to build fifteen margarine processing plants throughout the former USSR. In 1990 it gained another similar project for making food packaging materials which was worth approximately $6 million.

Summary

Trade with the former USSR has historically been a highly significant and financially rewarding aspect of this group's overall commercial activity. Hence, the necessity to react in a positive and constructive way to changing conditions in that country. On the one hand the changes in the USSR since 1985 have adversely affected the group's business. The admission of the dire economic condition of the USSR, and the consequent rethink of central government funding was a prime factor behind the Soviet decision to use their own gas turbines instead of purchasing Western products. This signalled the end of hitherto very profitable business for the power engineering company in the Soviet market. On the other hand, the new autonomy of individual enterprises has enabled the company to continue to benefit in other ways from its association with the USSR. The company felt that the formation of the joint ventures was a positive reaction to these changes and offered the prospect of continued profitability resulting from trade with the USSR. Moreover, the company believes that it is in a more favourable position to benefit from these new trading opportunities than other companies contemplating activity in the CIS market, because the joint ventures' end products can always be sold in Western markets to generate the hard currency necessary to make projects self-financing.

Case study no. 4 - a producer of cranes

Commercial relationships with the USSR and Eastern Europe in the 1960s and 1970s

In 1966 this company signed one of the first East-West industrial co-operation agreements with a Polish foreign trade organization responsible for the import and export of cranes and other handling equipment. In this agreement the UK company provided full design and manufacturing documentation for the production of a particular model of crane of thirty tons lifting capacity, and it supplied kits of items for use in the crane which the Polish enterprises were unable to manufacture. The Polish side delivered a programme of twenty-three items to the company including four sorts of chassis, gearboxes, slew rings and travel units to the company's technical specification, for subsequent assembly into various end items in the company's product range. These items were delivered to the company in mutually agreed quantities and at advantageous prices. The co-operation agreement was extended to include the sub-contracted production to Poland of twelve complete cranes of a modified version of the type covered by the agreement. This occurred at a time when the capacity of the British company was overloaded.

Various problems contributed to the lapse of the agreement in 1973. The main problem was caused by an action of the Polish side which was viewed by the UK company as not in accordance with the spirit of the agreement. The company had put much effort into developing a certain Middle Eastern market which was its exclusive territory under the terms of the agreement. Fifty-six cranes had been sold, operators had been trained and an after sales service had been established. On bidding for another 140 cranes in mid-1976, however, the company learned that the Polish side was also tendering for the same contract at prices which were approximately 50 per cent of those quoted by the UK company. The Polish side consequently won the contract and the increased demand on its production capacity meant that it was then unable to provide the chassis for the UK company. After this experience, no new agreements were signed subsequently with that organization; and the company formed the view that co-operation agreements in Eastern Europe should be restricted to product areas where the technology in the West was changing rapidly, to prevent unwanted future market competition from its Eastern co-operation partner.

Commercial relationships with the former USSR and Eastern Europe in the 1980s and the 1990s

The 1970s were good years for crane manufacturers with 22,000 cranes bought world-wide each year, but this number has now declined to around 6,000. This reduction in demand, coupled with increased competition from Japan, forced many crane manufacturers to cut back their operations. The company in question was consequently forced to reduce its numbers from 1600 employed in the 1970s to 200 currently, and it was incorporated into a larger company in the materials handling field in 1987. The smaller workforce reflects the reduced product range, following the decision to cease production of lattice boom cranes.

At the time of the 1966 industrial co-operation agreement the product range of the company included diesel-powered hydraulic mobile cranes, crawler-tracked cranes, rough terrain cranes, diesel powered mechanical mobile cranes, truck-mounted cranes and high quay cranes. Currently the company produces two main products - small hydraulic boom cranes and a new product based on crane technology launched in March this year - the rough terrain telescopic handler. This machine has taken over the traditional crane to a large extent - from its inauguration in 1974, annual world-wide sales have risen to 13,000. The changes in the company are also reflected in its mode of operation. It still conceives, designs, assembles and tests the cranes but no longer manufactures most of the components. Instead it is felt to be more cost effective to contract out for the majority of the parts.

In October 1990 the company signed a contract with the world's largest truck plant, Kamaz, based in the Russian republic 1000 km. east of Moscow. The Soviet plant had approached the UK-based Anglo-Soviet Trade Consultancy in order to find a suitable partner to enable it to convert its trucks into cranes as there was a distinct shortage of cranes in the USSR. As a company with previous experience in Eastern Europe, and one which exports approximately 70 per cent of its products, this UK manufacturer was receptive to the possibility of further expanding its overseas trade. Furthermore, the credentials of Kamaz as a potential trading partner were impressive. It had virtually no bank debt, was experienced in exporting its products and it employed a highly trained workforce. In 1990 its financial and organizational structure was formally changed to create a Western style corporation with a board of directors, an executive management structure and shareholders.

The first step of the agreement was to devise a way of securing the UK crane superstructure to the Soviet chassis. Therefore, the contract specified that the Soviet partner would provide a standard chassis on which the crane superstructures were to be mounted so that the UK partner could design an appropriate sub-frame on which to fix the crane. After this conversion is completed a Russian delegation will visit to assist with testing, and to undertake operation and service training on the new vehicle. The completed crane will then be shipped back to the former USSR where it will undergo safety testing before being accepted into the Russian road service. Following vehicle certification, the Russian partner is to begin production using kits to build the sub-frame and crane superstructures purchased from the UK partner. Engineers from the UK partner will be sent to the former USSR to help in the production of the sub-frames and in the assembly of the final product. At the time of writing, the agreement was in the early stages of operation with the Russian chassis at the UK plant and the prototype sub-frame almost complete.

The agreement is to cover twenty ton crane superstructures at the rate of 100 per year for an indefinite length of time, although some doubt was expressed as to whether this target would be met in practice. The terms of the agreement specified that the UK company could not sign a similar contract with any other Soviet manufacturer. Payment is in hard currency by letter of credit, against evidence of shipment. Kamaz was one of the companies given priority by the Soviet government to extend its business in Western markets and consequently it was allocated a source of hard currency from the central budget to achieve this aim. It was expected that payment for the superstructures would come from this source of funding and the British company was reasonably confident that these financial arrangements would prove to be satisfactory.

For the UK manufacturer the motives for entering this agreement were simply that it needed the orders for the superstructures. It is constantly looking for new markets and this agreement provided an entry into a country which is twenty years behind in some of its technical specifications for cranes. The UK partner is also trying to identify other opportunities for its products in the former USSR through an intermediary - the UK based Anglo-Soviet Trade Consultancy, the company which Kamaz had contacted initially.

The British company believed that the Soviet partner benefited from the agreement in a variety of ways as it has previously produced a large number of trucks for the armed forces, demand for which is currently being reduced. This agreement consequently enables the Russian partner to continue in the production of trucks but for a final product for which there is a market in the former USSR. The equivalent crane technology was not available in the former Soviet Union and the company's partner therefore needed to look further afield for a source of know-how. Finally, the partner is satisfying the former USSR's government pressure on Soviet industry to form closer business arrangements with the West.

The company found the representatives from the Russian partner to be very hospitable and enjoyed dealing with them. The only real criticism of the business conducted was the length of time taken at each stage of the negotiations, both in the time taken to secure agreement, and then in the protracted wait before the original Soviet chassis was sent over to the British factory. These delays are not yet over because the Russian side still needs to obtain approval for the final crane design from their authorities.

The UK partner is keen to establish further links with the new democracies of Eastern Europe and has had negotiations with a Hungarian organization which is looking for a joint manufacturing set-up. However, the company considers that there are still many problems in dealing with these countries, and scope for the development of their marketing and business skills for operating in a competitive international economy .

Case study no. 5 - a designer and manufacturer of mining equipment

Commercial relationships with the USSR and Eastern Europe in the 1960s and 1970s

This company's trade with Eastern Europe and the USSR during the 1960s and 1970s never exceeded 5 per cent of total exports annually. The contracts received usually included design, production and delivery of the equipment against the company's specification, followed by installation in an end-user's coal mine. The company signed three separate contracts with the Hungarian foreign trade enterprise responsible for the import and export of heavy engineering equipment in 1969, 1971 and 1973, for equipment for three separate coal mines. The company undertook the technical proposal and design work and the production of hydraulic and electrical equipment which was not manufactured in Hungary. The fabrications were produced by Hungarian enterprises, usually attached to the mining trust in which the equipment was to be installed. These fabrications were produced using the company's drawings, provided free of charge. The company also provided technological know-how where necessary and was responsible for supervising the assembly of the equipment in the Hungarian engineering works, and its subsequent installation in the Hungarian coal-mines.

Commercial relationships with the former USSR and Eastern Europe in the 1980s and 1990s

Although this company has been primarily a designer and manufacturer of mining equipment, it has expanded its product range since 1980 to include materials handling equipment through a process of acquisition and mergers. This has meant that the company can now offer a more complete package to its customers and does not need to sub-contract work out to the same degree as before. The company's workforce and its exporting activity have both been affected by the demise of the British coal industry since 1985. The workforce is currently around a third of the size it was in the 1980s, despite the fact that the company's product range has expanded. Furthermore, before 1985, the company was almost entirely dependent on the UK market for business, with exports amounting to around 10-15 per cent of turnover. Currently, the company's exports amount to more than half of its overall turnover. As exporting has become so important for the company, the former USSR and Eastern Europe have become proportionately more important markets.

Over the last ten years the company has received very little business from the East European and former Soviet market, and it is only in the last two to three years that it has made a conscious decision to invest more resources into developing this market. In 1991, the company obtained an order worth £0.75 million to supply the hydraulic equipment for a set of roof supports for a Hungarian coal mine. In a similar contract to those described above, the company provided the design which the Hungarian side used to manufacture the fabrications. A prototype of the end product is to be sent to the UK to be tested and approved by the company, before the Hungarian side can continue production at home. The company expects to receive payment in full through a letter of credit for its work in this contract. While the company is likely to gain on-going business from this mine for spares, it also hopes to gain other contracts from this coal-mine in the future.

This is the main contract the company has gained since making the decision to develop this market. The company felt that although the opening up of the USSR and Eastern Europe since 1985 had made it easier to establish contacts, it was no easier to gain contracts. However, the company recognized the long term potential of the market and was diverting considerable resources to increase its profile in the region. For example, the company's interests in Hungary have for many years been represented by the presence of an agent in that country. Recently, agents have also been appointed to identify and evaluate commercial opportunities for the company in Poland, Russia, Ukraine and Yugoslavia; and the possibility of appointing agents in Czechoslovakia, Romania and Albania is currently under consideration. These agents have succeeded in identifying potential business, although few definite contracts have materialized. This was attributed to the scarce availability of hard currency in Eastern Europe and the CIS, which encouraged their mines and companies to purchase the cheaper, less advanced mining equipment produced by a local manufacturer. The difficulty the company had experienced in gaining the necessary financial support to back possible projects was also discussed. Furthermore, the company felt that while it had really only started to consider the market seriously in the last two years, its German competitors, by virtue of their geographical proximity, had been building up contacts and achieving greater commercial involvement in this market for many years. It was hinted that this was likely to have produced greater awareness of German products in the market-place, giving them a competitive advantage.

However, the company is gradually building up its own contacts through attending exhibitions in Eastern Europe and the CIS and receiving visiting delegations in the UK. It decided to undertake extensive market research on each of the East European countries that were considered to offer the best potential for future business. This process has involved regular trips to the countries in question to meet with local manufacturers to discuss the potential for joint venture agreements, end users to evaluate both their product requirements and the resources at their disposal, and individual government departments which still have some influence. Hungary, Yugoslavia and Poland have all been researched in this way. Czechoslovakia is currently being investigated and Romania and Albania may be researched in the future. In Poland, a protocol of intent was signed with an existing Polish manufacturer in December 1991, which may lead to Polish production of equipment using designs specified by the UK company.

The company considers that the CIS, particularly Russia and Ukraine offer considerable potential for future business because of the insatiable demand for energy and associated equipment, which is unlikely to be met by local suppliers with the equipment available. Accordingly, the local production of the company's products is being explored, and, the company has signed a protocol agreement with the association representing the coal mining equipment companies in Ukraine, to initiate co-operation between the company and local manufacturers. Through the protocol, the company hopes to advise local manufacturers on ways to improve both their productivity levels and the reliability of their machinery, using designs which could be made locally with the addition of the company's technology and know-how.

Case study no. 6 - an industrial holding company/a manufacturer of transport equipment and associated components

Commercial relationships with the USSR and Eastern Europe in the 1960s and 1970s

The company described in this case study is a Midlands-based industrial holding company, which has had extensive experience of trading with Eastern Europe and the USSR, and at one point maintained a specialized office for the furtherance of East-West trade on a company wide basis. It signed two industrial co-operation agreements with East European organizations - one in the 1960s with a Czechoslovakian enterprise for the co-operative development of a special purpose casting machine, and another in the 1970s with a Bulgarian foreign trade organization for the production under licence of container handling equipment.

The co-operation agreement with a Czechoslovak enterprise This agreement, signed in the late 1960s, concerned the co-operative development of a special purpose casting machine. It involved the co-ordination of research and development, the exchange of scientific and technical information and the provision of technical assistance and know-how. No time limit was placed on the agreement as it was difficult to estimate the time needed to undertake the technical development. It was initially intended to extend the arrangement to a marketing agreement once the technical co-operation was successfully completed, but this option for commercial exploitation of the product did not materialize. This was because the Czechoslovakian partner appeared to be more interested in applying the technical development to its domestic needs, and also because the British company tended to view itself chiefly as a seller of parts and components, not specialized production equipment for their manufacture.

Currently, the company have no record of the current status of the agreement. It was the foundry equipment division of the company, a producer of sophisticated automatic sand casting equipment, which had signed the agreement, and this division has now been sold by the company. It was assumed by the executive interviewed that this signalled the end of the agreement as far as his company was concerned.

The co-operation agreement with a Bulgarian foreign trade organization In the 1970s, one of the company's manufacturing divisions produced a straddle carrier machine in response to the then world-wide demand for this type of equipment, caused by the containerization of all major ports. This division sold its machines world-wide and was responsible for signing a deal involving the Bulgarian firm, Balkancar - the largest manufacturer of forklift trucks in the world. This agreement, signed in the early 1970s, was to cover the following activities:

- a licensing agreement;
- provision of technical assistance and know-how;
- training of East European personnel;
- supply of parts and components to the East European partner;

- provision for parts and components by the East European partner for incorporation into the final product.

The Bulgarian partner paid an initial lump sum in sterling for the licence and associated know-how for the container handling equipment, and also purchased certain critical parts and components from the UK company. The company was obliged to buy back certain Bulgarian-made items for use in container handling equipment, and other materials handling equipment for distribution through its agency network.

As each port became containerized, the world demand for this type of large, expensive equipment reduced, and hence it was decided by the company to sell off this division. The new owners of this division have since closed the plant, making it difficult to ascertain the current status of the agreement.

Commercial relationships with the former USSR and Eastern Europe in the 1980s and 1990s

These previous links with Eastern Europe have not led directly to new ties being established by this particular company. Csepel, a Hungarian manufacturer of hydraulic power steering for commercial companies, offered a manufacturing licence to the company, but it was turned down as the product was not considered to be sufficiently advanced. The company remains interested in the potential demand for its products in the region and has maintained contacts of sorts with all the East European states and the former USSR. It believes that increasing its profile in the region is crucial if it is to benefit from possible business links in the future.

Since the signing of the previously described industrial co-operation agreements in the 1960s and 1970s, the structure of the company has not changed significantly, although a key individual in the company has since retired. This executive had been particularly well connected with personnel in the former USSR and Eastern Europe and was responsible for opening and operating the East-West Trade Bureau. His departure signalled the closure of this office and many of the company's contacts with Eastern European personnel were subsequently lost.

However, another company - a joint venture between this holding company and an American company, has been more actively involved in the former Soviet and East European markets and the remainder of this case study will concentrate on its activities. This company is a designer and manufacturer of transport equipment and automotive components, in particular trailer axles and suspensions. It also sells complementary products, produced by other firms, through its distribution network.

This transport equipment manufacturer's involvement in the Eastern European market started during the 1970s when Raba, a large Hungarian truck manufacturer, expressed interest in manufacturing the axles produced by the UK company. A dialogue was maintained between the two companies through the 1970s and 1980s, and Raba currently supplies the UK company with parts for its axles. The company is also currently selling its products to Hungary, Poland and Bulgaria indirectly through other customers. For example, it supplies equipment to a Swedish manufacturer which has formed a joint venture with a Polish enterprise. This joint venture company is now using parts produced by the UK company. Similarly, a Hungarian manufacturer is now also using its products by virtue of a deal with another Swedish enterprise.

The company has also endeavoured to make its products and name better known in Czechoslovakia, and executives have had meetings with the three top trailer manufacturers in that country. Currently, the company feels it has an opportunity to benefit from the recent changes in Eastern Europe affecting its industry. The aforementioned Raba used to be the key supplier of axles for Eastern bloc countries, but from January 1991 it decided to trade only in hard currency, making its products inaccessible to many of its previous customers. Hence, the UK company has started a dialogue with a Czech trailer manufacturer which has just begun production of axles, and has made proposals relating to this enterprise. It would like to separate the axle manufacturer from the trailer manufacturer to create an individual Czech axle manufacturing company which it could then control and run. The company hopes that its interest and intentions regarding this idea will be relayed to and considered by the relevant authority in Czechoslovakia, but it is pessimistic about the likelihood of a favourable response from the Czech authorities, because of the dramatic changes to manufacturing organization which the UK company considers would be necessary under its management. Despite the potential negative response expected, the long term objective of the UK company is to be the key supplier of the Czech trailer market, by filling the gap in the market caused by the changes in Raba.

In Bulgaria, discussions are underway with Somat, the biggest road transport fleet in Europe. All of Somat's trade is international and all of its vehicles are bought from the West; hence the UK company is aiming to supply its axles to the companies providing trailers for that organization. For example, Somat has recently formed a joint venture trailer manufacturing company with a German manufacturer and purchaser of the company's axles; hence the Bulgarian joint venture company will use the company's axles.

The company has also made its name known in Romania, albeit indirectly. The German manufacture referred to in the previous paragraph has also been involved in commercial activities in Romania which have resulted in the company's products being used by Romanian enterprises.

The UK company is also having an indirect influence in the Polish market through the strength of its position in Italy. Some Polish organizations have had long established trading links with certain Italian firms, some of which use components produced by the UK company. Furthermore, an agent was appointed in Poland in 1991 and the first orders for trailer running gear are currently being supplied.

Executives from the UK company have visited the former USSR in the last two years to evaluate opportunities in that market. Some of its products are being used there already as a trailer rental firm, which uses the company's equipment, has recently opened some depots in the former USSR. The company has also been involved indirectly in the former Soviet market through a counter-trade deal organized by an Irish counter-trade company. The object was to supply trailers in return for coal.

Unfortunately the deal has not been completed successfully at present, and negotiations are still continuing.

Case study no. 7 - a manufacturer of computer equipment

Commercial relationships with the USSR and Eastern Europe in the 1960s and 1970s

The company whose activities are described here, is a producer and developer of data entry and data collection equipment for use in distributed data processing systems. Originally the company was engaged in the production of flight simulators and large key-to-disc applications but it considered that the market potential for these products was limited. Consequently, the company purchased a licence for the production and marketing of the data entry and data collection equipment from its American designers in 1972 to enable it to enter a wider market place with a proven product. Around this time the company decided to develop the Eastern European market and hired a consultant for that purpose. Following a series of discussions with various Polish organizations concerning possible joint product development, it was decided to proceed with a licensing arrangement for the computer-related products in the company's range. It was signed with the Polish foreign trade organization responsible for the import and export of automation equipment and measuring instruments in 1974. The licence was to be implemented in a factory which formed part of the electronic measuring division of the industrial association responsible for the production of automation equipment and measuring instruments.

The licensing agreement commenced with an initial transfer of technology, followed by three distinct stages of product development. The agreement therefore included the following activities:

- the granting of a licence, though this was not sold as a separate item;
- provision of associated technical assistance;
- supply of associated components and materials;
- training of personnel;
- an agreement for marketing in specified geographic areas.

The main motives for the company's interest in the co-operation agreement were the expansion of market opportunities through exploitation of the licence, and the use of the licence to provide further business co-operation with the East European partner.

Commercial relationships with the former USSR and Eastern Europe in the 1980s and 1990s

The company now feels that the bankruptcy of the Polish economy in the late 1970s was the main reason for the collapse not only of the Polish licensing agreement business, but also the company's direct end user sales in Poland. It continued to supply a small amount of spare parts, but this dwindled to virtually nothing over the years.

In 1981, the company received a £7 million contract from the Soviet Ministry of Gas to supply seventy systems for the sub-stations in the Trans-Siberian East-West pipeline. This project was successfully completed but was considered to have been a great drain on the company's resources. It was felt that this factor may have been one of the reasons further contracts of this nature were slow to materialize from the USSR.

The company faced internal changes in 1984, when the holding company decided to sell off the various companies under its umbrella. The computer group to which this company belonged was then successful in a management buy-out bid. The new management endeavoured to continue its trade with the Eastern bloc and, in 1984, appointed an agent to market its products throughout Eastern Europe, with the exception of Czechoslovakia.

The company believes that the advent of the personal computer in 1984 was responsible for the subsequent expansion of its direct business in Czechoslovakia from 1985. There was a marked reduction in this business in 1989-1990 as a result of the political and economic upheavals experienced in Czechoslovakia, which made the company pessimistic about the short term prospects for sales of its equipment in that country. However, it has since gained two orders in 1991, valued at £330,000. The first was to provide the Bratislava-based trade organization, Koospol, which specializes in the import and export of food, with a computer system to handle the control of its commercial contracts as well as its accounting and administrative systems. The second contract involved the sale of three of the company's systems to the Czech State Insurance Company, a customer of the company since the mid-1970s.

The company has noticed changes in its business operations with Eastern Europe as a result of the region's democratization. As most of the familiar foreign trade organizations have either changed dramatically or ceased to exist, the company's main problem has been establishing the creditworthiness of the individual enterprises with which it must now deal. Furthermore, although the company was glad that the export licence parameters for some computer-based products have been relaxed,[4] it felt that this was counteracted by the tightening of hard currency expenditure. To compound these difficulties, competitors have entered the marketplace, some of whom offer cheaper products.

The company, therefore, has not benefited directly from its past agreements, but it has managed to profit in a small way from this existing user base with up-grades and add-ons. It is still interested in the long term potential of the East European market and is therefore monitoring developments closely.

Case study no. 8 - a nuclear engineering company

Commercial relationships with the USSR and Eastern Europe in the 1960s and 1970s

This company, whose main business operations were concerned with the design and manufacture of systems and equipment for civil nuclear reactors, was involved in exploratory work and joint conferences with almost all of the countries in Eastern Europe and the USSR in the 1970s. None of these proposals led to contracts, however, except for the sale in 1974-5 to Romania of a radio-isotope plant using the company's and the UK Atomic Energy Authority's (UKAEA) technology. This contract in turn led to the development of a joint venture between the Romanian side and the company, to be directed by two executives from the company, and two from the Romanian Institute of Physics. The joint venture was viewed by the company as a possible vehicle to enhance its prospects for future contracts in Romania; and, by the Romanian side as a vehicle for rapidly obtaining certain aspects of the company's technological know-how, and also as a means of selling Romanian engineering products in world markets.

In practice, however, it appears that these objectives were not achieved. The company has failed to gain any benefit in terms of future business from being part of a joint company, as illustrated when a contract for the project management and supply of equipment for a post-irradiation examination plant in Romania was awarded to a French competitor, after the UK company had invested a high level of resources at the detailed technical negotiation stage. Furthermore, the company was unable to find a market for Romanian nuclear equipment and other engineering products, whilst the Romanian side viewed some of the company's prices for control and instrumentation equipment to be uncompetitive. The Romanian side maintained that other business would materialize, but at the same time, unbeknown to the company, it was formalising a nuclear co-operation programme with the Canadian authorities. The joint venture company never did gain any business and the company now feels that its formation was the partial price that had to be paid in order to gain the original contract. The joint venture company is still in existence, but in name only. After it was informed about the Canadian deal in 1980, the UK company decided to regard the Romanian market as a low priority for the foreseeable future, considering it to be both unreliable in its methods of conducting business and lacking in adequate financial resources. Henceforth, the company decided to restrict its attempts to gain business in the region to other countries in Eastern Europe.

Commercial relationships in the former USSR and Eastern Europe in the 1980s and 1990s

In 1988 this company merged with a major French multi-product engineering corporation and in doing so considerably expanded its capabilities by establishing divisions in handling systems, marine and defence, power systems, manufacturing and construction, in addition to the nuclear division. It now carries out project management, engineering, planning, manufacture, assembly and test, site installation, commissioning and after sales service in all these divisions. The company is part of an international group which employs more than 80,000 people world-wide and makes over half of its £4.8bn. sales outside France and the UK.

The company feels that this merger has aided its prospects for trade with Eastern Europe and the former USSR, as it can take advantage of the many international contacts the group has. In each country in Eastern Europe, and in the CIS, the group employs local personnel who aim to identify market opportunities for the group's companies, and, who are subsequently in a better position to initiate contacts and negotiations.

The company is most optimistic about the future business opportunities in Czechoslovakia, Hungary and Poland and it is in these three countries that the more substantial contacts have been made. The company sees market opportunities for its nuclear division in these countries because the CIS is no longer amassing surplus fuel from nuclear power plants for use in the production of armaments.

Accordingly, the company has bid for a contract with the Hungarian electricity board, MVMT, currently in the throes of being privatized, involving the design and installation of spent nuclear fuel storage facilities, and feels confident that it will be awarded the contract. In order to assess the environmental implications of the company's proposals, it was asked by the Hungarian authorities to present and describe the system it would use to store the waste to an audience consisting of local councillors, environmentalists and politicians. As its particular method for waste storage is currently the one most favoured by ecologists, the company received acceptance for its proposals and felt grateful for being given the opportunity to present its case. Again, it was felt that the group's office in Hungary, staffed by local personnel, had aided the company's cause in finding out about and negotiating for this contract.

In October 1991, the company was involved in an almost identical contract in Czechoslovakia. It undertook a design study for a spent nuclear fuel store, for the Czechoslovak electricity authorities, CEZ. After being paid for this work, the company subsequently lost the contract to implement its design to a firm offering more competitive rates. The new contract has not yet been signed, however, and as it was not convinced that the competitor's designs would prove to be environmentally acceptable, the company hopes that it may still gain the business.

The company felt that its attempts to secure a manufacturing base in Czechoslovakia in which to produce its designs for the storage of the spent fuel, illustrated some of the problems inherent in conducting business with these new market economies. It was advised by the Czechoslovak power board to consult the Czech engineering company, Skoda, regarding its proposals for manufacturing the company's product. Consequently, a protocol agreement was drawn up between both companies which should have led to a binding contract. However, Skoda had decided to collaborate with an American company and therefore informed the UK company that henceforth it would be carrying out all its nuclear work with its new US partner. The UK company returned to the Czech authorities and was advised to go to the large Vitcovice engineering firm in Ostrava. Again, a contract was drawn up, but, after the UK company had submitted its final proposal with this firm, it learnt that a competitor, offering a different type of technology for this job, had signed an agreement with another part of Vitcovice. The company's view that these instances highlighted the underdeveloped commercial awareness of Eastern bloc companies was strengthened when it discovered that Skoda had decided to forgo its co-operation with the American firm in preference for a new agreement with a German company.

The company has also responded to an enquiry for a baggage handling system for Bratislava airport in Czechoslovakia, involving the supply, design and supervision of the production of the parts. It is expected that a large factory situated near to Bratislava airport could undertake the production to the company's design. This factory was originally built to service military planes but has experienced a marked reduction in this work and has therefore had to diversify into other areas to continue production - for example it is currently making trailers for lorries.

Case study no. 9 - a marketing consultancy

Commercial relationships between the USSR and Eastern Europe in the 1960s and 1970s

This company is a small marketing consultancy operating predominantly in the agricultural sector but with a growing range of industrial clients. It is run by a UK consultant who has now diversified its operations by opening bases in Holland, Hungary and Czechoslovakia and by employing an associate in each.

This consultancy was established in 1981 out of what was previously a designer and manufacturer of equipment for use in the agricultural industry. This predecessor had been active in trading in many of the Eastern European countries, selling a number of products to each of the socialist countries, and acting as contractors for the provision of know-how for two package deals delivered to buyers in the German Democratic Republic. Most notably it had signed a ten-year industrial co-operation agreement with the Hungarian 'Komplex' foreign trade association in 1968.

This agreement involved the transfer of technology for continuous oil fired grain driers of the company's advanced design, in which the Hungarian partner purchased a licence to manufacture the equipment, and associated technical assistance to facilitate production. The company also offered to train some Hungarian personnel, and agreed to provide up-to-date technology over the time of the agreement. The Hungarian partner had rights to market the equipment in all of the countries of Eastern Europe and in certain other countries with the company's permission. This agreement expired in 1978 and as there was nothing to renew, the co-operation between the two partners ended. The company felt that the Hungarians had kept to the agreement meticulously and that the agreement had operated successfully. One problem was that other Hungarian organizations may have copied the technology but the company felt that there was little which could have been done to prevent this.

One of the main motives of the company in securing this agreement was to enable it to remain in the Hungarian market. It had previously sold grain dryers and other equipment to Hungary but had reached a point where further sales were impossible because of currency restrictions, although the buyer still maintained a high degree of interest in the company's products. The agreement failed to circumvent these problems in terms of future business for the company in the Hungarian market. However, it was felt that this was due to inherent shortcomings in the Hungarian economy rather than any failings in the agreement. Furthermore, it was only three years after the agreement had ended that the company transformed the nature of its business from a manufacturer to a consultancy.

Commercial relationships with the former USSR and Eastern Europe in the 1980s and 1990s

From 1981, when the present consultancy was established, until 1990 it ignored the many approaches received from Eastern Europe. These approaches generally were made by companies with no money to spend and who therefore required a two-way barter arrangement. The goods they produced were generally of inferior quality and were already available in the UK. The consultant felt that the previous Komplex deal had been very export oriented and in this sense was quite unique. With no quality goods, or goods for which there was insufficient demand in the West, there was therefore little opportunity for East/West agreements to be made in this sector.

In 1990, however, this consultancy responded to the changing political and economic situation in Eastern Europe by opening two bases in Eastern Europe - one in Prague and one in Budapest and appointing an associate in each. These bases were set up to offer marketing services to those companies who want to export to the West. The consultants in the Eastern European offices establish the nature of the business required by their clients and the research is undertaken by either the UK consultant or the other associate based in the office in Holland (covering the Belgian, Danish and German markets) as appropriate. Similarly, the research acquired in the UK for contracts concerning the Eastern European markets is undertaken by the associates based in those countries.

For East European companies contemplating the move into Western markets, the consultants initially aim to establish whether the company's products are viable in the Western country by:

- assessing the demand in the market-place for that product, and,
- establishing the nature of the competition in the country - both indigenous and imported - in terms of product range, pricing structure and marketing methods.

After this stage the consultant comes to a conclusion as to whether there is a market for the Eastern European product or not. If it is decided that a market exists for the product then the consultants offer a further service to develop the strategy to be undertaken. They can supervise the marketing effort by finding distributors in the country and by acting as the company's representative in the market-place.

So far the Hungarian base has proved to be financially rewarding with six figure orders being taken in hard currency. Since May 1991, in particular, the amount of business achieved has far exceeded expectations. In Czechoslovakia, however, the pace has been slower. This was blamed primarily on the lack of hard currency and the economic situation, but it was also felt that the underlying lack of trust in the middle man created by years of over-bureaucratization was another explanation. Furthermore, the consultant felt that many businessmen in these countries were unaware of the factors that should be taken into consideration when marketing a product or service. The idea of market research, for example, is still a relatively new one. In order to explain and promote the services of the consultancy a circular is being sent out, emphasising that a well researched and thought out marketing strategy is essential for the success of any product in Western markets. They stress that employing a specialist to provide this service is more reliable, thorough and cost effective than the company attempting to do it for itself. Furthermore, the consultancy aims to provide a cheaper, smaller scale alternative to the major firms performing these services already in Hungary and Czechoslovakia.

The UK consultant felt that he had found a niche in the market-place through the setting up of these bases. He believed that the service they aimed to provide was one which was badly needed and would help the Eastern Europeans acquire a more Westernized approach to doing business at a more affordable price than the alternative large-scale consultancy firms. He said that he was very fortunate to have found the two associates who are currently running the offices in both Hungary and Czechoslovakia. The Czech was an ex-head of the Czechoslovak Institute of Agricultural Engineering and the Hungarian a graduate in agricultural engineering. Without local personnel to buy and run the office space the UK consultant felt it would have taken far longer to set up the operations, been fraught with legal difficulties and would have placed his consultancy in a less favourable position to gain local business.

The UK consultant was adopting a wait-and-see attitude regarding the East European bases and their future. Although more than satisfied by the response so far, particularly in Hungary, he felt it was too early to predict what the future holds in these countries.

Case study no. 10 - an aero-engine manufacturer

Commercial relationships with the USSR and Eastern Europe in the 1960s and 1970s

This company, as a designer and producer of aircraft engines for use in both civil and military applications, signed two licensing agreements with Romanian foreign trade organizations. This case study charts the subsequent development of these agreements and the progress of discussions with other prospective CIS partners. The company also purchased a licence for an investment casting plant during the 1960s, and the features of this agreement are discussed in chapter 5.

In 1972 a licensing agreement was signed with both the Romanian Technoimportexport foreign trade organization and the Ministry of the Machine Building Industry, for an engine with military applications. The agreement covered a licence to manufacture the engine, engine assembly, and engine overhaul. A similar licence was also sold to a Yugoslav foreign trade organization, and the Romanian and Yugoslav partners intended to share production of the engine components which were then to be exchanged so that both sides could assemble the engine, to be installed in a particular model of aircraft of joint Romanian/Yugoslav design. The intention was for the licensees to move towards complete self-build of the engine, with the licence being purchased on an initial lump sum basis, with a royalty payment for each aero-engine manufactured.

This agreement is still continuing although its final aim of complete self-build of the engine has not yet been achieved. The production of a range of engine components has been implemented, but these are being sold to the UK company instead of being assembled to build the engine.

A further agreement was signed in 1979 with the Technoimportexport foreign trade organization and the National Aeronautical 'Centrala' (CNIAR). The licensee purchased the right to manufacture a significant proportion of the components, and to assemble the engine. The remaining components were to be purchased from the company including proprietary supplied items. The engine was then to be fitted into the airframe of the BAC 1-11 which was to be manufactured under licence in Romania - the Rombac 1-11 (see chapter 4). The licence was purchased on an initial lump sum basis, with a programme of purchase of components until 1985, and the agreement was due to continue until 1999. Credit facilities for the customer were arranged by a consortium of banks with Export Credit Guarantee Department (ECGD) support, and provided at a relatively low interest rate. As in the 1972 agreement, an obligation for counter-trade was incurred, with a bias towards aero-engines and other products of the Ministry of Machine Building Industry.

Commercial relationships with the former USSR and Eastern Europe in the 1980s and 1990s

A variety of difficulties materialized in the operation of this 1979 agreement which have resulted in its standstill, the main problem being financial. The low aircraft production rate of the Romanian side was the prime factor behind its inability to meet its financial obligations, as it barely produced one aircraft a year and failed to sell the completed product.

One of the main reasons for the Romanian partner's failure to sell the few aircraft that it produced was that the technology of the engines was superseded, as the engine covered by the agreement was replaced by another engine in the late 1980s. The original engines cannot now be sold, because the current Western international noise regulations, which were introduced earlier than expected, specify requirements which cannot be met by that engine. These problems came to a head in 1987-8 when the

UK company refused to provide the technical surveillance of the Romanian factory which was necessary in support of any recognition of the engine by the CAA. This decision was prompted by the Romanian side's refusal to make the additional payments for the technical surveillance by the UK company which had been so conditioned. The Romanian partner still has ten complete engine kits which it would like to resell to the UK company for sale as spare parts.

The future of this agreement is not promising. The Romanian side has for many years been eager to take up the new engine in a similar type of agreement to that for the original engine. Whilst the UK company had initially agreed to convert the licence for the original to a licence for the new engine, the Romanian side's failure to meet its existing financial obligations led to the withdrawal of its offer, although it undertook to supply complete new engines in support of a re-engined Rombac programme.

The main motive for the British company when entering the agreement was to obtain income from the sale of licences and from the ensuing trade in engines, components and other goods in a market where it was difficult to sell finished aero-engines. Furthermore, it hoped that the agreement would provide the company with the opportunity to develop a long term relationship with Romania. Having analysed the development of the agreement it seems that each of these aims were only partly achieved through no apparent lack of effort by the UK company.

This company is still predominantly a designer and producer of aero-engines, but the acquisition of another group of companies in 1988-89 enabled its capabilities to be increased to include design, manufacture, commissioning and in-service support expertise in power engineering, nuclear engineering and general mechanical engineering. These industries form around 30 per cent of the company's business. They already have some small-scale ties with the former USSR and are looking to extend their role further in the former USSR and Eastern Europe. The other 70 per cent of the business, which formed the original agreements, is receptive to developments in the USSR and Eastern Europe and is interested in becoming involved where possible in these countries. The remainder of this case study considers the various attempts at co-operation by the aero-engine division of this company.

As yet the UK company has not signed any new agreements with these countries, but there are various options being considered. The main change as far as the aircraft industry in the USSR and Eastern Europe is concerned is in its structure. Previously the aircraft industry was controlled by the state, with five separate organizations performing the various areas of aircraft production in the USSR which are usually undertaken by one company in the UK. Because the industry was sheltered from competition and fluctuations in consumer and market demand, no marketing research, sales or promotional skills were required in any of these organizations, which were mainly preoccupied with fulfilling their quota demanded by the state. As the transition to a market economy gets underway, the role and structure of these separate organizations is changing dramatically with each either withering away or aiming to act in a similar way to a Western company. Many of the design bureaux, for example, are seeking to expand their roles through contact with the West.

Currently there is close co-ordination between the UK company and the Tupolev Design Bureau in Moscow concerning an agreement to install the UK company's RB211-535F5 engines into the Soviet Tu-204 aircraft. Two engines are due to be sent out in 1992 to be installed into an aircraft which will also be fitted with Western avionics and interior. At the time of researching this case, it was hoped that the completed aircraft - similar to the Boeing 757 - would be demonstrated at the Farnborough air show in September 1992 (this subsequently occurred), and approved to Western certification standards by 1994. The UK company hopes that this co-ordination will eventually result in the formation of a joint venture agreement, although it envisages that the design bureau will merge with the aircraft manufacturer before an agreement is signed with the company and Western financiers. The involvement of a Western marketing company is also anticipated to effect the sale of the aircraft to Western markets to gain hard currency. Encouraged by the prospects for this agreement, the company is currently discussing the feasibility of similar schemes with other design bureaux.

Co-operation with the USSR is also anticipated to materialize as a result of the formation of a new airline, 'Air Russia' - a joint venture company between British Airways, Aeroflot and the Ministry of Civil Aviation. The agreement was signed in August 1991 and it is planned to start operating in 1993 from the Domodedovo airport in Moscow, which is undergoing refurbishment as part of the deal. It will operate internationally to gain Western currency, initially using the Boeing 767-300 aircraft. The UK company expects that these aeroplanes will be powered by its engines, and is enthusiastic about the Air Russia development because this airline should manage to overcome the usual problem of a lack of hard currency.

The UK company cited other deals and agreements of this nature and is generally optimistic about its future trade with the USSR and Eastern Europe. To ensure it will have every opportunity to be included in prospective deals, it is closely monitoring the actions of the leasing companies supplying these countries and is also researching the aircraft needs of the leasing companies with the aim of selling their engines to them. Leasing offers the most affordable option for the USSR and Eastern Europe to achieve the incorporation of Western technology. For example, in Bulgaria the Balkan airline company has begun leasing the A 320 aircraft. This is fitted with a joint venture engine with partners from the US, Italy, Germany and Japan, as well as the UK company.

Comments

Some of the material contained in this chapter reinforces, to a certain extent, some of the conclusions presented in chapter 2 above. In the first place, those companies which had originally sold licences in the 1960s and 1970s for technology related to the automotive and agricultural engineering industries in Poland and Romania, did not receive any follow-on business related to those commercial arrangements during the 1980s and 1990s. As in the case of Soviet investment in the automotive industries discussed in chapter 2 above, therefore, the East European investment in these sectors during the 1960s and 1970s appears to have been a one-off large scale investment to increase capacity in those sectors, with little subsequent supporting investment. A similarity with some of the companies reported in chapter 2 is also revealed by the radical changes taking place in their internal organizational structures and ownership during the 1980s and 1990s. One company, for example, which had sold a shock absorber licence to Romania sold its operations to a Belgian company, whilst another company which had sold licences to Poland, closed down its centralized office responsible for the establishment of trade contacts in Eastern Europe.

Of the ten companies whose activities are described above, however, six companies have maintained their trading ties in the region. It is difficult to establish the extent to which the experience of doing previous business in the USSR and Eastern Europe during the 1960s and 1970s had ensured their continued presence in the market, as most had found the trading conditions prevailing currently to be so different to those existing under the old system. Only one company, for example (case study no. 5) signed a contract in the 1990s which resembled the one it signed in the 1970s, but the competences which companies would have acquired through their previous trading contacts and experience and in industrial co-operation, would have been the following:

- knowledge of the structure and composition of the Eastern bloc foreign trade organizations;
- established contacts with key personnel in the relevant foreign trade organizations;
- knowledge of the bureaucratic structures and processes and an approximate expectation of the time needed to negotiate an agreement;
- an understanding of the state central funding system and an expectation of eventual payment through that source;
- an understanding of the need for countertrade, and an ability to link these payments to exports of products, components and technologies from the British company.

The opening of the Eastern bloc has altered trading conditions so profoundly however, that some of the advantages of prior knowledge of the market and its trading conditions appear to have been negated. In the case of one company, for example (case study no. 8), the nature of the demise of its Romanian agreement encouraged the company to neglect that market for the foreseeable future. In the case of another company with many years of experience in the market (case study no. 2), well established trade contacts with reliably paying foreign trade organizations were being put at risk as decentralization progressed.

The experiences of several companies in this market have been far from negative, however, and point to definite positive signs for success in the new transitional conditions operating in the former Soviet and East European economies. In one case, for example, a crane manufacturing company which had been far from satisfied with its previous experiences of industrial co-operation in Poland was in the process of adapting its products to be installed on a Russian truck (case study no. 4). This vehicle was produced in a factory which appeared to have sufficient reserves of hard currency to cover the costs of design, testing and component purchase. The Russian factory had also been engaged in the production of trucks for military purposes, and would consequently benefit from the allocation of funds for conversion to civilian manufacture as the demand for military vehicles was reduced. In the case of another company engaged in the manufacture of automotive transmission units (case study no. 6), a large potential market was beginning to emerge in Eastern Europe as the former Comecon trade partners began to change to trade in hard currency, thus making some Western products more financially cost-effective than previously.

In other cases, companies appear to have been successful when they have been able to relate their sales to products for which there appears to be a definite market potential in the West. The aero-engine company, for example, (case study no. 10) after having experienced many difficulties in its previous trading relationships with Romania, was hoping for improved commercial contacts with the former Soviet Union through the sale of engines adapted to fit to an airframe with sales potential in Western markets. In the case of the power and chemical engineering group of companies (case study no. 3), the business potential in the former USSR has radically improved through the establishment of a set of joint ventures to refurbish CIS plants, and sell their products in Western markets; although the traditional power engineering business of the company had suffered a temporary setback.

In both of these apparently successful cases, therefore, the companies appear to have capitalized on existing Russian technological capabilities where their partners appear to have potential for using those capabilities in a cost effective fashion. The realization of these potentials has required the outlay of high levels of corporate resource, and the ability to develop and construct the appropriate financial frameworks to support the required directions of business.

Notes

1. Hill, M.R., *Export Marketing of Capital Goods to the Socialist Countries of Eastern Europe,* Wilton Publications, Farnborough, 1978, pp. 179-204, 209-13.
2. Hill, M.R., *East-West Trade, Industrial Co-operation and Technology Transfer,* Gower, Aldershot, 1983, pp. 75-101.
3. Hill, M.R. - 'Les relations economiques URSS-Royaume Uni' in *Le courrier des pays de l'est*, November 1990, no. 354, p. 52
4. Hill, M.R. *Soviet Advanced Manufacturing Technology and Western Export Control,* Avebury, Aldershot, 1991.

4 Large scale industrial co-operation projects

Introduction

This chapter describes the activities of three companies who have been partners in large scale industrial co-operation agreements with organizations in Eastern Europe. These activities were similar in principal to those described in the previous chapter, but were more extensive and complex. These three industrial co-operation agreements were originally signed during the 1970s, and related to one major industry in each of two Eastern European countries, namely the aeronautical engineering industry in Romania and the agricultural engineering industry in Poland.[1]

In one of these companies, however, it was found that industrial co-operation with the Polish partner had ended in 1982, as the company ceased trading during that year. The company concerned (GKN Contractors Ltd.) was a large project management and engineering organization, involved in two technology transfer arrangements related to the manufacture of tractors at the 'Ursus' factory. Contracts were signed in 1974 for the Polish purchase of forging and casting technology, and a further contract was signed in 1976 for the purchase of bimetallic bearing technology. The business arrangement for the Ursus project consisted of a single 'software' contract to supply management and know-how services over a 51 month time interval for three plants. The company considered that it obtained this contract for two main reasons: it was well-known to the purchasing Polish foreign trade organization, 'Metalexport' with whom it had previously signed a contract for the supply of equipment to the Jawor forge; furthermore, it was a major supplier of forged parts to Massey Ferguson in the UK , for the same range of tractors to be produced under licence in the Ursus factory. It was consequently recommended by that company in its role as major technology supplier to the Ursus group of factories.

The remaining two cases relating to the Romanian aeronautical engineering industry and the Polish agricultural engineering industry, are described below.

Case study no. 1 - British Aerospace plc

Relationship with Romania in the 1960s and 1970s

On 31st. May 1979, following intense negotiation, an industrial co-operation agreement was signed between British Aerospace plc (BAe) and the 'Tehnoimport' foreign trade organization of Romania, for the production under licence in that country of the BAC 1-11 500 and 475 series aircraft. Although collaborative relationships have now become almost commonplace amongst aircraft manufacturers, this was the first case of a jet transporter ever to have been licensed in its totality from one national manufacturer to another. This case study records the subsequent development of this agreement.

The forerunner of British Aerospace plc before nationalization took place on 29 April 1977, was the British Aircraft Corporation (BAC). The so-called 'Rombac' agreement emanated from many years of trade between BAC and the Eastern bloc.[2] Some seventeen years before, the company had sold three second-hand Vickers Viscounts to the 'Motoimport' foreign trade organization for use by the Polish national airline. The value of this sale was some £1.4 million, which was entirely covered by means of countertrade in Polish ham. As far back as 1962, therefore, the company had gained important experience in countertrade activity, which was to prove beneficial in future dealings with its Romanian partner.

The Romanian business interest commenced in 1967 when BAC became aware of the desire of Tehnoimport to acquire Western aircraft. In May 1968 it was agreed to sell six BAC 1-11's and spares to Romania after the company had managed to organize payment for these aircraft through a framework agreement which linked together several independent contracts between different principals, both buying and selling. The total contract price of just over £10 million was to be about 80 per cent covered by the Romanian side through the exports of machine tools to an agreed British agent; the assembly and delivery of Islander aircraft to the Britten-Norman company on the Isle of Wight using materials, components and assemblies provided by that company; and through the export of other products to a Swiss trading company. In 1975 another contract was signed by both parties for five aircraft with spares, and in the late 1970s the Rombac deal was eventually signed after the initial inquiry by the Romanian side in October 1976, concerning the possibility of purchasing a licence to manufacture the entire airframe of the BAC 1-11.

That agreement was signed by both parties with the following aims in mind:

Production levels It was hoped that by the end of 1986 the Romanian side would have built a total of twenty-two aircraft from kits supplied by BAe, and would have become self sufficient and able to continue production independently, apart from the requirements for vendor parts.[3] Forty of the aircraft expected to be produced were designated for Romania's own use, but CNA (the National Aeronautical Centrala, which was the successor to

Tehnoimport), also placed emphasis on exporting the product at the earliest opportunity in order to earn foreign exchange.[4]

Financial arrangement The agreement, valued in excess of £150 million to BAe, was to be totally covered by hard currency payments from Romania via a long term credit facility arranged by UK banks. The British position was underwritten by ECGD. BAe also entered into a countertrade agreement with the Romanian government to identify buyers for a wide range of Romanian products in markets throughout the world. Of the £33 million countertrade to be created at then current values, 20 per cent was to be taken from the aircraft industry, partly as 1-11 components for spares, but mostly in the production of more 'Islander' aircraft; 60 per cent was to be taken from other engineering industries responsible to the Ministry of Machine Building Industries, and the final 20 per cent was to be taken in other commodities.[5]

Expectations of the Romanian partner in signing the agreement It was hoped that the deal would provide the Romanian tourist and scheduled markets with a modern jet. The Romanian side was proud of the aviation industry which had existed in Romania prior to the First World War,[6] and looked towards the agreement as a means by which it would be able to recreate a modern aviation industry through co-operation with a Western partner.

Expectations of BAe in signing the agreement Prior to nationalization in 1977, BAC felt that with the BAC 1-11 it had the best medium haul airframe in the world with which to develop new models for the 1980s. The aircraft at that time was fitted with a Spey engine, but until the company could be certain that the engine manufacturer would develop and deliver a new generation of quieter, more fuel-efficient engines, BAC needed also to consider the option of using engines from other manufacturers. The belief that it also needed a cheap manufacturing source with whom to collaborate in the development of these new 1-11 variants was a motivating factor behind the Rombac deal.

Furthermore, BAC's initial interest in a potential agreement with the Romanian partner was in accordance with its long term view that collaborative ventures were an effective means of protecting and expanding its role in the market-place. It was hoped by personnel at BAC that the arrangement with Romania would eventually evolve into a joint marketing arrangement involving both its expertise in selling, and cheaper Romanian labour to complete the product. BAC was also attracted to the prospect of penetrating Third World markets by utilising the system of bilateral trading between governments, then prevalent between members of the Council for Mutual Economic Assistance. A final and highly important factor increasing the potential attractiveness of this venture, was that BAC expected to receive a large hard currency payment of £13.5m for the sale of the licence.[7]

The operation of the agreement The company organized thoroughly for the job, and a huge logistic operation was arranged. All major and proprietary parts collected at Hurn Airport were taken away in a variety of specialized large capacity cargo aircraft arranged by BAC, together with Soviet aircraft sent over by the Romanian partner. The latter party then began to assemble, build, fly and certify the aeroplanes, and on 23 March 1983 it celebrated the inaugural flight of its first home-produced Rombac 1-11.

In practice, however, the agreement failed to match up to the expectations of both parties involved. The Romanian side continued to build the aircraft, but at the disappointing rate of around one a year which it would exhibit at the annual air show at Farnborough or Paris. All twenty-two kits were delivered but only ten aircraft were built. Structural parts of the remaining kits were unused and other components were used as replacement exchange items to supplement Romanian production or 1-11's operated worldwide. The low production levels attained by the Romanian side were blamed on infrastructural problems inherent in the Romanian system under President Ceaucescu, which even prevented adequate supplies of fuel and power to the factories.

Similarly, the desire of the Romanian side to export the finished product to Western markets proved to be unrealistic in practice. The Romanian side had no sales team or experience with the idea of selling, and failed to comprehend the techniques required to succeed in the competitive world of commercial aviation.[8] These disadvantages were compounded because the aeroplane itself had a limited market in the West due to the fact that the Spey engine, assembled under a separate licence and fitted to the aircraft, was ceasing to become acceptable under Western legislation because of its noise levels. Although each aeroplane was fitted with hush kits, this factor did not aid the Romanian side's attempts to sell to Western markets. Furthermore, the new generations of aeroplanes which arrived in the market-place in the 1980s led to the Rombac 1-11, built to 1970s specifications, becoming relatively old fashioned in terms of cabin style, airworthiness requirements and avionics.

The Romanian side was consequently forced to adapt its strategy of sales to Western markets in recognition of these difficulties. The completed aircraft were sold to Tarom, the Romanian national airline, which was given the responsibility of leasing them to Western markets to generate at least some hard currency. This move proved to be reasonably successful after the appointment of a new leasing manager who understood Western aviation. He succeeded in leasing aircraft to the UK and, to two airlines in Yugoslavia, among others. Tarom, meanwhile, continued to use Soviet aeroplanes as they could be accounted for in more easily accessible 'soft' roubles.

The initial hopes for collaborative development and joint marketing also failed to emerge as BAC itself underwent changes through the nationalization process, which were felt by some to have profoundly affected the development of the agreement.

The process of nationalization involved the reluctant amalgamation of BAC and Hawker-Siddeley,[9] two companies, which up to this point had been competitors in the same industry. A period of uncertainty ensued whilst personnel were appointed to take over the management of the new company. Eventually it was decided that the newly formed Commercial Aviation Division of British Aerospace would be run chiefly by personnel recruited from Hawker-Siddeley. This decision led to the departure of some key individuals who had been instrumental in setting up the original Rombac deal and who were eager to continue co-operation and collaboration with the Romanian side. The new personnel at BAe Commercial endeavoured to finance the re-launch of the Hawker-Siddeley 146, which was a four-engined, six abreast, high wing regional jet serving similar markets to the BAC 1-11. In July 1978, in one of the two main decisions of the nationalization period, BAe received £250 million of government funds to finance the relaunch of the BAe 146.[10] According to a viewpoint expressed to the authors by a previous BAC executive involved in setting up the Rombac agreement, priorities at BAe shifted perceptively from the BAC 1-11 to the BAe 146 at that time, and the potential Romanian production of the BAC 1-11 began to be regarded in terms of competition with the BAe 146. Hence, although BAe continued the supply of all kits as contracted, when this was completed the company reduced supplies to, and contact with, Romania. This, it was felt by the former BAC executive, effectively extinguished all the initial hopes for collaborative development and joint marketing with the Romanian side that were fostered by BAC up to the time that the definitive protocol (the precursor to the formal agreement) was signed at ministerial level on 28 May 1977; and is viewed by that executive as one of the main reasons why the Romanian desire to re-engine the 1-11 with the new Tay engine has failed to materialize to date.[11]

An alternative viewpoint, expressed by current BAe personnel, is that nationalization had little effect on the operation of the agreement. These personnel emphasized that BAe had performed its obligations to the Romanian side meticulously, but that full payment for the twenty-two kits of aircraft parts supplied was never received, despite the fact that over 85 per cent of the countertrade negotiated was also achieved. It was hinted that it was this factor that was likely to have hindered further co-operation, rather than any other.

It is difficult to assess the effect of nationalization on the agreement because of the evident production problems in Romania which severely affected the operation of the agreement and also because BAe remained in public hands for only four years. With the return to power of a Conservative Government in May 1979, BAe underwent another change in structure and personnel when it was sold off in two instalments between February 1981 and May 1985. This did not result in any immediate re-consideration of the Romanian deal, however, as the initial concerns of the new management were with the projects it had inherited, and not with reviving old agreements.

Plans to resuscitate the agreement[12] There have been several reports since 1986 of attempts to rekindle and extend various aspects of the Rombac manufacturing potential. Business interest has been maintained because the BAC 1-11 has proved to be a long lasting executive aeroplane, and the installation of new high capacity engines could give further life to an effective airframe design having a good business potential, particularly in the aircraft leasing market.

A further factor influencing this business interest was the surplus of BAC 1-11 400 series aircraft in the United States during the mid to late 1980s. This surplus provided consequent potential for the absorption of design, development and certification costs, which were not expected to be exorbitant, over a number of saleable modified aircraft which could be purchased at a competitive price. In addition, successful developments on the 400 series may have been transferrable to the 500 series aircraft. If certification was successfully achieved for these developments, a licence for the manufacture of these designs could be sold to the Romanian manufacturer, which could then be used as a cost-effective source of supply. Attempts at further business developments of the BAC 1-11 and Rombac agreements have consequently focused on the refitting of existing BAC 1-11 airframes with Tay engines, and the potential for using the Romanian factory as a cost-effective source of supply for modified new aircraft.

As illustrated by the reports cited in subsequent paragraphs, however, most of these plans have encountered a series of problems. Some of these problems were apparently due to cash flow requirements which are a common feature of aeronautical engineering research and development expenditure; and the securing of financial resources to meet these expenditure requirements has not been assisted by the limited capability of the Romanian manufacturer to contribute hard currency for technical development. Furthermore, it is likely that the UK airframe and engine manufacturers would have been cautious in committing themselves too deeply to the extension of a venture which had experienced previous problems of production output and consequent financial contribution, and which may have demanded resources which could have been engaged on other more profitable and strategically important projects. Finally, if the surplus of other second-hand narrow bodied passenger jet aeroplanes became too great, there may not have been a profitable market for the re-engined BAC 1-11 aircraft.

In August 1990 it was reported[13] that a British aircraft leasing company (Associated Aerospace) was to acquire fifty BAe 1-11's from Romania with an option to take another fifty, in a deal worth an estimated $1bn. over the following eight years. In order to achieve this aim, which relied on the aeroplanes being produced by the Romanian manufacturer, (Interprinderea de Avioane Bucuresti) at a rate of ten per year by 1994, the co-operation of BAe, Rolls Royce (who were to supply the Tay 650 engine), and the AIM group (to supply interior kits), would have been required. The length of time spent attempting to secure agreement apparently took its toll of time and resources, however; as that proposed project did not come to fruition.

An American company, Dee Howard of San Antonio, Texas, owned by Alenia, an Italian electronics firm, has also been engaged in efforts to re-engine the BAC 1-11. In December 1991, however, it was reported[14] that this programme had been halted temporarily and that one hundred staff employed on the project had been laid off. The problems emerged because Dee Howard had experienced difficulties in gaining support for the supplemental type certificate. The company was believed to have spent close to $80 million on the project without achieving the necessary support from certification authorities and suppliers. The report also suggested that a large surplus of narrow bodied passenger jets existed, which may also have reduced the market attractiveness for re-engined BAC 1-11s.

Throughout the 1980s BAe had kept a permanent presence in Bucharest with quality control personnel monitoring and supervising the Romanian production line. Furthermore, one of the key companies involved in the countertrade deal set up by BAe in the 1970s, (Britten-Norman) had managed to survive two receiverships and its successor, Pilatus-Britten-Norman Ltd., was still importing around forty Islanders a year from Romania. Moreover, the new generation of Romanian politicians were eager to do business. Thus, all the main parties involved in the original relationship were still in existence.

The newly appointed Managing Director of Commercial Aircraft Ltd. in BAe was favourably disposed to the idea of continued co-operation with the Romanian side; a new design and engineering group was formed to handle the Tay Rombac project, and, with the halting of the proposed Associated Aerospace project referred to above, BAe began to wonder whether it could not resuscitate relations itself.

BAe devised an idea for a new aeroplane, the BAC 1-11 2000 variant, composed of the Rombac 1-11 with new Tay 650 engines and a new computerized flight deck and interior designed by AIM. However, it has taken some time to translate ideas into reality and although BAe signed an agreement on 6 March 1990 to pursue the plan for an upgraded BAC 1-11, there are still doubts as to whether the project will materialize. The new flight deck and interior would be very costly which prevents BAe from making a hasty decision. Furthermore, continuing fears that the production in Romania could take orders away from its BAe 146 'whispering jet' programme, by offering a cut-price alternative, has reportedly made BAe reluctant to grant the Romanian government permission to licence an updated version of the twenty eight year old BAC 1-11 design.[15]

The financial difficulties associated with the operation of the initial agreement may also have dampened enthusiasm on the restart of the programme. It was reported[16] that this new BAC 1-11 concept is so well regarded by the European Bank for Reconstruction and Development (EBRD) that it may be prepared to help fund the programme if BAe or another manufacturer were prepared to back it. However, BAe did not expect that financial support from bodies like the EBRD would be readily forthcoming, although it was admitted that a strong case could be made relating to the social consequences which would be felt in Romania if the agreement collapsed.

In November 1991 it was reported[17] that the Romanian government's plans to resurrect its aircraft industry had won the backing of Federal Express, the US freight company which operates the world's seventh largest airline. It has shown strong interest in placing orders for up to thirty two aircraft with Romania. This may put intense pressure on BAe to give the necessary design permission to modify the airframe, although the

UK company still has reservations concerning the ability of the Romanian side to produce aircraft at a commercially viable rate.

An alternative strategy also being considered by BAe management is that there is still a market for the Rombac 1-11 with the Spey engine in economically underdeveloped parts of the world, where noise regulations are not in effect. This idea emanates from the belief that the airframe is still impressively strong and robust - 'the DC3 of the post-war period'. As a result of the fragmentation of the Soviet Union, it was also felt that the richer republics like Georgia and Ukraine would need to operate cross border flights independently of Russia, and would prefer to spend money on a cheaper aircraft with a sound reputation. Hence, the aircraft could be sold as a replacement for the ageing Soviet-built Tupolev 134 currently used by Eastern European airlines.

At present, therefore, the agreement is at a stalemate stage where the argument is finely balanced between a wasted opportunity of letting the agreement wither away, and the risks and pitfalls associated with restarting production.

Commercial relationships with the former USSR and Eastern Europe (other than Romania) in the 1980s and 1990s.

BAe and the variety of companies under its umbrella have been unable to use the Romanian agreement as a platform for the development of new links in that country to any great extent. Although sales have been made of some minor components and many contacts have been established, the main problems associated with Romania twenty years ago apply to an even greater extent today, and are not conducive to the development of trade. There is little hard currency to purchase goods and there are few opportunities for countertrade because of the poor quality of most Romanian goods.

In the former USSR, however, some tangible agreements have been made. In August 1991, the Rover car company (a subsidiary of British Aerospace), launched its Montego and Maestro models in the Soviet market.[18] The agreement, worth up to £50 million to Rover, was organized through Orbicom, a UK based trading company, who received the original order for 4,800 cars from the Siberian regional government. As part of the deal Rover agreed to modify all the cars to specifications which take into account the severe climatic conditions and uneven roads in Siberia. Orbicom is to act as the barter house in the deal and will purchase liquid natural gas from the Siberian regional government and sell it for hard currency in the West in order to pay Rover for their cars. Orbicom also signed a protocol with NAMI, the state-controlled organization responsible for both vehicle technology and design in the USSR and the monitoring of standards of imported cars, which provided for the eventual establishment of a Rover test centre in Moscow and the basis for a retail distribution network.

Subsequent negotiations are reportedly underway between Rover and NAMI, to set up plants to produce Rover cars in the new Commonwealth of Independent States.[19] In an attempt to capitalize on low labour costs and to be in a position to take advantage of market opportunities within the CIS and Eastern Europe, Rover is considering proposals to start production of its established Montego design in Russia. Sites capable of building up to 150,000 cars a year are being examined in central Russia and Siberia, but at the time of writing, no decision has been made. BAe is likely to be offered a direct equity stake in NAMI, which BAe may also request be privatized. The potential deal involves Russian assembly of British made components and may eventually result in NAMI aiding in the modification of Rover designs. The cars will be available in roubles and it is hoped that they will be purchased by Russian entrepreneurs and from firms who have benefited from the transition to a market economy. It is uncertain how the deal is to be financed, although it was hinted that imports could also be paid for through the sale of oil. It was emphasized that while discussions with BAe are the most advanced, NAMI is also considering other potential partners from Europe and the Far East.

Another area of co-operation between BAe and the USSR was announced at the International Air Show in Farnborough in September 1990. The idea behind this collaboration is to use a Soviet aircraft, the Antonov 225, to launch BAe's new low cost reusable space vehicle, HOTOL. The continued discussions on this programme may eventually end in a joint venture agreement. This is an area of co-operation which would have been inconceivable even five years ago because of the obvious link to military applications. In the current international climate, however, there are fewer such restrictions and BAe is attempting to take advantage of one area of Soviet expertise - a highly advanced space industry, resulting from the Cold War of the preceding decades.

Other areas for possible co-operation lie in environmental clean-up operations in badly polluted areas, particularly in Eastern Europe, for which EC funding is likely to be available. Furthermore, BAe sees the new political system operating in the region as an opportunity for co-operation on defence matters. All of the states of Eastern Europe have individual security fears and will attempt to build up their own defences as soon as they can afford to. They are likely to turn to a Western defence company for ideas and supplies and may therefore present a market opportunity for the company.

Case study no. 3 - Massey-Ferguson-Perkins Ltd - a manufacturer of agricultural equipment

Commercial relationships with the USSR and Eastern Europe in the 1960s and the 1970s

Massey-Ferguson-Perkins Ltd. was a wholly-owned subsidiary of Massey-Ferguson Ltd, and the operating company which entered a large scale industrial co-operation programme with the Polish tractor industry. The main agreements were signed with the Agromet-Motoimport foreign trade organization during October 1974 and formed the largest single trade and industrial co-operation agreement between a UK company and an East European socialist country at that time.

The agreement related to the modernization of the Polish tractor and diesel engine industries focusing on the extension of the 'Ursus' group of tractor factories, based in Warsaw, by the introduction of tractors and engines in the 38-60 HP category. The complete co-operation programme was divided into four main sections namely:

- a general agreement, outlining the objectives of the agreement and acting as the framework for the remaining sections of the programme;
- a licensing agreement relating to the product range for which licences were to be purchased and trade marks were to be granted;
- an industrial co-operation agreement relating to the sale of completely knocked down (CKD) tractors (to be subsequently assembled in Poland) and other components to Poland, and the ensuing sale of products to Massey-Ferguson-Perkins;
- an implementation agreement which specified the planning, supervisory and consultancy responsibilities of Massey-Ferguson-Perkins.

The following activities were carried out during the course of the agreement:

(i) Purchase of licences These purchases related to not only the traditional made-in components of the tractor, but also those elements of a tractor that were usually bought in by the company, namely brakes, clutches, gaskets and other components and commodity items which usually accounted for some 20 per cent by value of a total tractor. Massey-Ferguson-Perkins consequently purchased licensing rights from the suppliers of these items on behalf of their Polish partner. It was unnecessary to obtain licences which had been purchased separately by the Polish partner or where existing Polish technology was deemed sufficient to produce the product (eg. for plastic and rubber moulding processes, and bearings and pistons).

(ii) The provision of associated technical assistance This related to the provision of complete technical documentation for material specifications, process planning and associated tooling, plant layout and factory services, quality control systems and production control systems. This documentation involved all items covered by the industrial co-operation agreement.

(iii) The purchase of associated components and materials The initial phase of the tractor build programme was achieved with the company providing the Polish partner with part-knocked-down tractors which were then assembled in Warsaw. By 1981 the Polish partner was producing 35 per cent of the tractor components, with the remaining 65 per cent being delivered by Massey-Ferguson-Perkins. The final objective was for the Polish partner to produce some 98 per cent of the tractor components.

The company entered a counter-purchase agreement to buy some £165 million of Polish produced items - a sum based on the total investment value of the project. Some of these products could be sold by the company through its distribution network, and some could be used in its own factories.

(iv) The exercise of quality control by the company The company's quality and validation procedures formed part of the technology transfer. The company also retained the right to carry out any cross checking within the enterprise, to ensure that the counter-purchased products were interchangeable with those supplies drawn from other sources.

(v) The provision of training The provision of training by the company was extensive initially, with teams of personnel from the Polish partner being trained at the company's training centre and in sub-licensee's establishments. Furthermore, Massey-Ferguson-Perkins located a team of four engineers in Poland to assist during the inaugural phase of the agreement and an office was established in Warsaw to undertake correspondence between the UK company and the Polish side through a telex link. Another office, located in the UK, was staffed by Ursus and Agromet personnel.

While the technical involvement of Massey-Ferguson-Perkins declined as the agreement progressed, there is still a company business manager working on the agreement in Poland and the executive interviewed continues to visit Poland around six times a year to monitor developments.

(vi) An agreement for marketing in specific geographical areas The Polish partner was given the right to sell licensed tractors throughout all of the CMEA countries and Mongolia. Outside this region, however, it was only allowed to sell to the company.

The development of the agreement 1974 - 1991

This section will aim to assess to what extent the original aims of the agreement were fulfilled.

Production levels It was hoped that the production levels of the plant could be expanded from 50,000 Ursus tractors per year to 75,000 MF tractors and 90,000 engines, and a further 20,000 Ursus tractors in the '60 HP plus' class per year, over a fifteen year period from 1974. Furthermore, it was hoped that the Polish partner would eventually be producing 98 per cent of the total tractor components.

The highest volume of tractor output achieved, however, only approached 60,000 per year with a workforce of 15,000. (This compared with the 80,000 produced by 5,000 employees at the company's plant in Coventry). During the late 1970s and throughout the 1980s Ursus was selling 40,000-50,000 MF tractors on average each year in the Polish domestic market. In 1991 the number of MF tractors that Ursus produced and sold fell to 15,000-20,000 tractors.

However, the aim for complete production of the tractors at Ursus was achieved by the early 1980s, although until recently Massey-Ferguson-Perkins was still providing parts worth up to £1 million a year, as a result of certain production problems, or to top up levels of certain components.

Financial arrangement The value of the Polish purchases of licences, associated technology and manufacturing equipment was estimated at some £127 million at 1974 prices. A loan for that sum was arranged by Barclays Bank International on behalf of a syndicate of eight British banks to the Bank Handlowy w Warszawie with ECGD guarantees for £104.45 million. It was intended that the deal would be completely self-financing from the Polish foreign currency viewpoint as Massey-Ferguson-Perkins agreed to purchase some £165 million of Polish-made products.

The UK company received the money for the licence from the Polish side, but the buy-back deal negotiated in 1974 was severely affected by Western inflation levels in the late 1970s and Massey-Ferguson-Perkins consequently agreed to increase its buy-back commitment to £310 million. However, in 1986 the two sides signed a deal which altered the nature of the arrangement. In return for reducing the buy back commitment to £6.7 million, the Polish side was given the go-ahead to export the Massey-Ferguson-Perkins designed tractor to other, previously forbidden markets, though not under the company's trademarks.

This deal suited both sides for a variety of reasons. It became evident to both sides that Massey-Ferguson-Perkins would find it extremely difficult to achieve the buy-back target as the decline in world demand for tractors meant that it was having enough difficulty in selling its own products. Massey-Ferguson-Perkins had intended to achieve the revised buy-back target by buying components made by its Polish partner, but in reality the UK company has spent less than £1m on Polish parts and components. The real business has come from buying the cheaper, less sophisticated Polish final product and selling it in the North American market to week-end farmers with low usage requirements. Through these purchases Massey-Ferguson-Perkins has more than met its amended buy-back obligation. Ursus itself has also exported around 2,000-3,000 of the Massey-Ferguson licensed tractors to overseas markets including Greece, Scandinavia, Ireland and Africa.

The aims of Massey-Ferguson-Perkins in entering the agreement

These aims can be summarized as follows.

- The company was eager to expand production and diversify into new markets as a result of the surge in world demand for agricultural machinery in the 1970s. It already had some experience in the Eastern European market by virtue of a 10 year licensing agreement with Yugoslavia, signed in 1955, and through its negotiations with a prospective Romanian partner in 1972. The role of supplier of products, components and technology seemed to represent the best means of entry into the favoured Polish market in view of foreign currency restrictions on the purchase of finished products in that country.
- Massey-Ferguson-Perkins realized that an acceptable volume of income could be attained in this venture from the initial sale of the licence and the subsequent sale of CKD tractors and components.
- It felt that it could expand its international market opportunities on a competitive basis by having access to another supply source where no direct capital investment was required.

These three main aims of Massey-Ferguson-Perkins were broadly achieved in practice. Although the agreement has been beset by various problems during the last fifteen years, most have been more detrimental to Ursus than the UK partner. The agreement did provide the company with an entry into a new market and ensured that its name was better known throughout Eastern Europe as a result of the association. It also received an initial large sum for the licence and more recently has benefited from its ability to source a cheaper product in Poland.

The aims of the Polish side in entering the agreement

The Polish partner's reasons for entering the agreement stemmed from national requirements to modernize agricultural production and become self-sufficient in food. As one of the largest tractor factories in the world, Ursus aimed to modernize its product range of agricultural equipment, expand the volume of tractor output and establish the Polish agricultural engineering industry on a more technological footing.

The company felt that the Polish partner had broadly achieved these aims through the agreement. Poland is now self sufficient in food and has improved the technology of her agriculture. Although the anticipated production levels failed to materialize and Poland's agriculture remains under-mechanized, this was blamed on the lack of a sound financing system in Poland to encourage retailing and the general economic malaise prevailing in that country.

The Ursus plant, though, was besieged with other problems which culminated in the declaration of its bankruptcy in 1991 causing strains in the Massey-Ferguson-Perkins - Ursus accord. Some of the problems were external and arose from the financial problems associated with Western inflation and the effects of the Western recession which provoked a slump in the world tractor market. However, Massey-Ferguson-Perkins felt that some shortcomings in the management of both resources and people was the main problem contributing to the financial plight of Ursus. Recently, company personnel have gone to Ursus to work on the shop floor to check the quality of the products to be sold in the US. The overall impression was of a factory which was overmanned, with a constantly changing labour force, which paid little attention to quality and did not co-ordinate activities on the shop floor.

Commercial relationships with the USSR and Eastern Europe in the 1980s and 1990s

Status of the agreement - October 1991 In July 1991 the Ursus factory was declared bankrupt with debts of $203 million. The Director General of Ursus and the Polish Minister for Industry were both replaced and the newly appointed Minister for Industry set up an internal consultancy team to study the operations at Ursus and come up with recommendations regarding its future viability and survival. It announced its statement on 2 October 1991 which recommended that the Ursus factory be saved following a shut down for five weeks, presumably to reorganize the plant. 2,500 immediate redundancies were announced with the total expected to lose employment as a result of the reorganization estimated to be approximately 6,000-8,000.

This recent statement was seen as a positive development by the company's management who believed there was a possibility that the Polish government would declare Ursus unviable and too expensive to save. Furthermore, it was felt that a rationalization process would prove beneficial to the future operation of Ursus. The company had considered overmanning to be one of the prime reasons behind Ursus' problems as the total workforce of the Warsaw site was some 13,000, with 1000 people employed in the marketing department alone.

Since the end of 1989 and throughout 1990-91, the company has been selling the MF 231 tractor, built in Poland, to North America. To date over 3,000 tractors have been sold in this market. The Israeli market was also targeted for these tractors but product quality problems have restricted sales so far to only twenty-four units. Among the other markets considered has been Mexico.

However, early in 1991, Massey-Ferguson-Perkins began to meet with reluctance from Ursus in relation to its intentions in the Mexican market, including a refusal by the Polish side to give price quotations. This apparent lack of co-operation aroused suspicions at Massey-Ferguson-Perkins, which were subsequently justified by the news of the activities of 'Art B', a Polish holding company. Art B had agreed to buy the entire tractor production at Ursus in April, and thereafter around 3,000 tractors per month, for cash. It had taken over all of Ursus's domestic distribution and had started to exert influence in some of its export markets as well. One of Art B's target markets was, in fact, Mexico (the other was Argentina). Art B's offer for the sale and local assembly of tractors, including MF licensed tractors, was reported in the Mexican press, provoking indignation at Massey-Ferguson-Perkins and prompting immediate negotiations between the company, Ursus, Art B and Ursus Trade, the joint venture company set up between Art B and Ursus. Fraught discussions continued between May and July 1991 and Massey-Ferguson-Perkins were at the point of withdrawing from the licence agreement altogether.

The discovery at this point that Ursus was on the verge of bankruptcy and that Art B's financial status was also questionable, forced Massey-Ferguson-Perkins to take stock of the situation. Two of the three founders of Art B departed to Israel when it became evident that their financial operations up to this point may have been bordering on the illegal.[20] As Art B's credibility was undermined there was a dispute about who owned the 7,000 tractors that Art B had supposedly purchased - the bank or Art B? Ursus Trade was wound up and negotiations were re-established with Massey-Ferguson-Perkins. A matter of concern for the company was that BHZ Ursus, the foreign trade office, would have ceased activity during this time as its duties had been reduced. This would have meant that Massey-Ferguson-Perkins would have lost a great many useful contacts who could influence future co-operation. Fortunately this office was restored giving the UK partner grounds for optimism regarding its future ties with Ursus. Moreover, the company still believes that technically Ursus is a sound and well-equipped production base and this provides an added incentive for continued co-operation.

Other trading links with the former USSR and Eastern Europe

The company is eager to benefit from the new opportunities that it feels have materialized in the former USSR and Eastern Europe in the last few years. It believes that it must have an advantage in this changed market-place because of its experience in Poland and the benefits that this has given it particularly in terms of product awareness.

To date, however, the Polish connection has failed to provide it with any other concrete trading links in the region largely because of the severe financial difficulties prevailing there. The company is visiting and researching these countries regularly to be in a position to take advantage of any trading opportunity as it arises. It believes the former USSR to be the market with the most potential for future trade because of its sheer size and agricultural requirements.

Comments

The information presented in this chapter has illustrated the high levels of corporate skill required to develop and implement large scale industrial co-operation agreements, and many of these aspects of corporate skill were becoming apparent in those successful companies described in chapters 2 and 3.

A major factor emerging from these case studies, however, is the high degree of risk existing in this type of large scale transaction if internal conditions in the East European partner's country fail to support the intended business development. The industrial infrastructure in Romania failed to support the ambitions of the Rombac agreement, particularly in terms of output rates for the 1-11 aircraft. These low output rates had a consequent effect on the generation of foreign currency, although this problem was further compounded by a lack of East European expertise in Western markets. It is apparent, therefore, that there are still many lessons to be learned from this case study, which will also have some relevance to the company's major suppliers of aero-engines, avionics and aircraft interiors. British Aerospace now retains some links with the previous Romanian partner, but now views the USSR as a more promising market, particularly as the company can now offer a broader product range to include automobiles as well as aircraft, following its recent acquisitions. In the case of Massey Ferguson, the continued operation of the industrial co-operation agreement depends on the continued existence of the Ursus factory, which in turn depends on the factory's capability to compete in a time of turbulent national-economic transition. These factors are clearly beyond the company's control.

Notes

1. Hill, M.R., *East-West Trade, Industrial Co-operation and Technology Transfer,* Gower, Aldershot, 1983, pp. 102-32.
2. Hill (1983) pp. 102-10.
3. Hill (1983) p. 113, and *Flight International* 15 March 1980, pp. 852-4.
4. Hill (1983) p. 113.
5. Hill (1983) p. 112.
6. *Touchdown (undated)* - (journal of the BAC Commercial Aircraft Division) no. 3, pp. 2-20. ('Pioneer in World Aviation' by D. Clarke).
7. Hill (1983) p. 111.
8. *Flight International*, 12 March 1983, p. 655.
9. See Hayward, K., *The British Aircraft Industry*, Manchester University Press, Manchester, 1989, pp. 133-88; Gardner, C., *British Aircraft Corporation*, Batsford, London, 1981, pp. 278-88
10. Hayward (1989) pp. 153, 157-8.
11. *Flight International*, 12 March 1983, p. 657.
12. The information for this subsection of the chapter was obtained from an interview with the previous executive of BAC who was consulted for the first part of the chapter. This executive was involved with the BAC negotiations with Romania which culminated in the 1979 licensing agreement, and he has continued to maintain a close interest in that project.
13. *International Trade Finance*, 9 August 1990.
14. *Flight International*, 11-17 December 1991, pp. 4-5.
15. *The Guardian*, 28 September 1991, p. 10.
16. *Ibid.*
17. *The Guardian*, 12 November 1991.
18. *The Financial Times*, 6 August 1991, p. 3.
19. *The Guardian*, 14 January 1992.
20. An account of the operations of 'Art B' was published in the colour supplement of *The Observer,* 4 October 1992, ('Zloty Zillionaires').

5 Licence purchases

Introduction

This chapter describes a sample of five companies which all purchased licences from the socialist countries of Eastern Europe in the 1960s and the 1970s. The estimates that were made at that time, of the volume of trade in licences and associated payments between the socialist countries of Eastern Europe and the developed Western economies, suggested a predominantly eastward flow of licences and associated technological know-how, and a predominantly westward flow of payments either in hard currency or goods. Hence, the majority of published research on East-West technology transfer, was chiefly concerned with the eastern flow of western technology; and studies of the westward flow of eastern technology were consequently carried out by one of the present authors,[1] as that topic had not been widely researched.

The activities of the following companies were therefore investigated to provide information on the westward flow of Soviet and East European technology at the company level. The subsequent development of these licence agreements is recorded below, for the following companies:

- the chemical division of a steel company;
- a manufacturer of rechargeable batteries;
- an iron foundry;
- a plant engineering company;
- a foundry in the aerospace industry.

The chemical division of the steel company purchased a Bulgarian licence for the coating of electrodes used in electric arc furnaces for the production of steel. One of the company's engineers had previously read an article in a European steel journal on a Bulgarian process for electrode coating, and arrangements were subsequently made for discussions with the Bulgarian foreign trade organization responsible for licence exports. This resulted in the company being granted an exclusive licence for the UK in 1968.

Following the purchase of the original patent, the company improved the process and product operating parameters to enable substantial savings to be made in electrode consumption, and consequently managed to secure for coating some 85 per cent of the total of 17,000 tons of graphite electrodes used annually in the UK in the late 1970s. In addition, the company also purchased an exclusive licence from the Bulgarian licensors, for exploitation of the licence in Sweden in 1974, and negotiated non-exclusive licences for Canada, USA, France/Benelux and West Germany.

The steel company sold off this coating process in June 1991. Another company was formed to take-over this process, but it declined to be interviewed on the grounds that it now has no links with the business communities in Eastern Europe or the former Soviet Union.

The manufacturer and distributor of rechargeable batteries purchased a licence from, and was involved in a joint venture with, a Hungarian organization engaged in the development and production of medical equipment, which also had the right to engage in foreign trade. The original joint venture was set up in the late 1960s, but the Hungarian partner's share was purchased by the company in 1980, when a tighter licence agreement was signed and the company was granted sole agency rights in the UK for the product that was the subject of the agreement, namely sealed zinc rechargeable cells. This licence lapsed in 1985-6, and as the delivery and quality of the components delivered from Hungary had become variable, the company did not attempt to prolong the agreement.

The iron foundry was part of a large and diverse engineering group which purchased a Soviet licence for a foundry process, relating to the fluid sand process, in which moulding sand was mixed with a liquid binder, a self hardening chemical and a binding agent. This process differed from conventional casting methods, since the sand was not compacted around a pattern to achieve the requisite cavity shape, but was simply poured and subsequently hardened by means of the chemical reaction. The licensing agreement was signed in 1968 and covered the sale of the licence, associated technical assistance, and exclusive rights to the UK with the possibility of sub-licensing to other British foundries. This process was particularly successful for the casting of large ingot moulds, bottom plates, ladles and balance weights, and as a result of its successful application, the company has managed to sub-licence the process to another British foundry. The factory using this process could not now be traced however, as it has been closed down following re-organization of the parent company, and the present authors could find no evidence of the licence being used elsewhere in the UK.

The remainder of this chapter describes the activities of those two companies which have retained some contact with the former Soviet and East European market, following the purchase of licences from that source around twenty years ago.

Case study no. 1 - a plant engineering company

This case study relates to the experiences of three sister companies operating in the metals division of a large UK corporation which provides project management, design, manufacturing and construction services through three divisions to the world's metals, process and off-shore industries.

Introduction

The iron and steel company forming part of this metals division purchased a Soviet licence for the cooling of blast furnaces in 1972. The licence was purchased from the Soviet foreign trade organization responsible for the sale of licences and was due to run for 10 years. Apart from other reasons relating to Soviet technological expertise in the use of stave cooling furnaces, conclusion of the agreement was facilitated because of long standing trading relationships between the USSR and other companies within the metals division and other divisions within the corporation. These included the sale, by another company within the metals division, of an ore pelletizing plant to the USSR in the late 1960s and the subsequent sale of methanol plants and forging presses. Consequently, the licence arrangement was viewed as one element of a long term business relationship between the corporation and the USSR. This licence however, was only used on four occasions and will terminate shortly. The support of this technology from the Soviet side has also diminished greatly over the last few years. The corporation's links with the former USSR are still intact, however, as evidenced by the continued presence of an office in Moscow and the activities of two other companies in the metals division of this corporation, as described in more detail below. Trade with the former USSR and Eastern Europe amounted to 2.2 per cent of the overall turnover of the metals division in 1990-91.

Company 1 - a designer, builder and installer of continuous casting machines for billets, blooms and slabs

In 1986 this company was acquired from the British Steel Corporation to provide the group with the capability of supplying a fully comprehensive range of plant and equipment to iron and steel industries throughout the world. This company within the metals division is currently the most active in the corporation in forging trading links with the USSR and Eastern Europe.

Commercial relationships with the USSR and Eastern Europe in the 1960s and 1970s During the period when it was still part of the British Steel Corporation, this company was more active in modernizing the UK and other Western steel industries than considering other business opportunities abroad. One bid was submitted for an order for a caster in a new Polish steel plant, but a competitor won the contract. By the 1980s, however, most Western and UK steel works had adopted continuous casting processes and a need developed to consider opportunities in other markets. The USSR and Eastern European markets were targets because of the lower level of technical development of their steel industries.

Technical change in world steel over the last three decades has been associated with the introduction of oxygen steelmaking and continuous casting. Since 1960 the oxygen converter has rapidly displaced the open hearth furnace as the primary process in the industrialized West; and, whereas the old processes required the reheating of the ingot for primary rolling, continuous casting directly provides semi-finished billets or slabs suitable for subsequent hot working operations. Soviet industry showed an early awareness of the significance of these two major post war innovations and in the latter case it played a prominent role in the development of the process,[2] but since the late 1960s the diffusion of both oxygen steel and continuous casting has been much lower in the USSR than in industrially advanced capitalist countries.[3] Whereas the construction of open hearth furnaces ceased in Japan in the late 1950s and the USA by the mid-1960s, the Soviet industry was introducing new capacity as late as 1970. This decision reflected the existence in the USSR of both an efficient open hearth technology and also the ready availability of relatively low cost scrap which tended to make the traditional method of steelmaking more economic, as scrap cannot be used as a material input to the same extent in the oxygen process.[4] The slow rate of the introduction of oxygen steel making capacity had consequent effects on the diffusion of continuous casting facilities as the oxygen converter is the most suitable installation for continuous casting. This was also the case in Eastern Europe in 1988 with Poland, Hungary and Czechoslovakia producing around 40 per cent of their steel output in open hearth furnaces, and with only an estimated 10 per cent of the Polish and Czechoslovakian final output being continuously cast.[5]

Because commitment to full employment was implicit in the centralized planning system and the operation of the Comecon co-ordinated system of international trade, there was little incentive or motivation for Eastern bloc states to make production more labour efficient by streamlining their operations. Intra-bloc trade exempted these countries from competition in world markets and ensured they would always be able to barter their steel for another Eastern European commodity, albeit of lower technological sophistication than that available on Western markets. There was consequently an absence of market pressure to improve product quality or increase volume of output, through major process innovations.

Commercial relationships with the former USSR and Eastern Europe in the 1980s and 1990s From the company's perspective this lower level of technological development was an encouraging feature of the market-place as it meant that there were fewer plants installed with continuous casting equipment. Furthermore, both the former USSR and Eastern Europe were felt to be rich in manpower and the natural resources required for steelmaking.

The wave of change which has swept Eastern Europe and the former USSR since the late 1980s has meant that the old certainties have disappeared, and there is now a real need for all former Soviet and East European enterprises to earn foreign exchange and to become profitable and viable. In order for steel enterprises to achieve this aim it is generally considered that they need to undergo dramatic rationalization in terms of manpower and machinery in order to produce steel to Western standards.

Although the world demand for steel has reached a plateau, the company has felt for some time that Eastern Europe and the USSR represented long term consumer markets for steel products. The company is currently attempting to secure contracts to undertake the modernization of two steel plants in the former USSR - the Karaganda steel plant in Kazakhstan and the Lipetsk steel plant in Russia. These two were selected after the company had undertaken a lengthy study to assess the potential of all steel plants in the former USSR, and a description of each site is given below.

1. Karaganda: This is a large integrated iron and steelworks built in the late 1930s to exploit local coal and iron ore resources. It is currently producing more than five million tonnes of steel per annum (this represents 40 per cent of the total output of British Steel) with no continuous casting facilities. Until a recent fire in the tinning line the plant produced 40 per cent of all Soviet tinplate. It was selected for further development for the following reasons:

- Size: The size of Karaganda was important - both in terms of the established plant and the extent of modernization required. The company believed that the steelmaking practices at Karaganda were in need of updating and that a continuous casting facility should be installed if it was to compete in world markets in the future. Currently 320 tonne converters are struggling to produce quality steels using the outdated ingot technique. The total cost of modernization of the entire facilities was estimated to be £400m, with about half (£186m) of this total being required to pay for the process plant to be imported.

- Export experience: Karaganda already has had important experience in the sale of some of its products, most notably its hot rolled coil products, for hard currency to Western markets.

- Situation: Karaganda was also chosen because of its position in the Republic of Kazakhstan. The company felt this was an important consideration because of the increased autonomy of the individual republics in the former USSR. Kazakhstan was the second largest republic in the former USSR in terms of land area, and was felt to have considerable resources and potential. It holds the largest known untapped oil-field in the world and is well endowed with natural resources including gas, copper, gold, zinc, coal and lead.

- Project overview: The development of the steel plant calls for the complete reorganization of the existing shop, downstream of the converters with new desulphurization stations, new tilt and rabble facilities, new secondary steelmaking units and new vacuum degassing facilities. The existing shops are to be extended by two bays to house the new slab continuous casting machines. All the new equipment and processes will be supplied by the company with full supporting services and full training of the customer's personnel to run the plant.

  By any standards this proposition is a major steelworks development and would increase production and improve the quality of the

Karaganda output to six million tons of steel of a quality suitable for the world market.

- Timescale: Karaganda have major site clearance problems which it expects to take 24-30 months to solve. To finalise civil engineering plans, approximately half of the process equipment design will need to be completed at an early stage. The following timescale is envisaged.

October 1991:	Site work commenced
Early 1992:	Preliminary design order
Late 1992:	Full commitment order
1994:	Commence plant installation
1996:	Commissioning

- Financial overview: Deciding on the type of financial arrangement to underpin the agreement is proving to be the most exacting task surrounding the entire agreement. The UK company felt that at present many Western banks are reluctant to finance commercial transactions with the former USSR and Eastern Europe and that this was a severe obstacle to the UK's trade with that market. This unwillingness was blamed on the current economic recession and the recent experiences of banks in their transactions with the Third World. It was felt that if the opening up of the former USSR and Eastern Europe had taken place twenty years ago, finance would not be the restricting problem it is today.

As stated previously, the total development cost is currently estimated at £400m, of which £186m relates to the cost of the process plant to be imported. The balance of £214m is expected to be a rouble expenditure for building, civil engineering and associated services. It is hoped that the project can be financed through an escrow account although attempts are still being made to find finance through banks and financial institutions. Karaganda has an established source of hard currency with the sale of its products through the TSK steel organization and agreement in principle has been reached with the Kazakhstan government to finance the project via this route.

TSK Steel is a joint venture company which was formed in March 1989 between the Karaganda Steel Complex (69 per cent) and SYTCO, an international trading company based in Switzerland (31 per cent), to be the vehicle to export some of Karaganda's products and purchase its equipment. Since its formation, TSK has handled all of Karaganda's equipment purchases and it is intended that it will also purchase the tinning line and the casters.

An escrow account has the advantage that no loan is required to set up the agreement. It is hoped that payments to the escrow account would be made from the export sale of pig iron, hot rolled coil, tin-plate, and ultimately slabs, in hard currency through the TSK-SYTCO route. Delivery of products is to be guaranteed by Karaganda and the Kazakhstan government. If Karaganda cannot meet its quotas then pre-planned alternative sources of supply will be provided by the Kazakhstan government. If Karaganda sells its products through other channels, then the necessary hard currency values will be transferred to SYTCO to cover payment commitments. If the agreed financial arrangement requires a guarantee, then SYTCO would give a 'running' $400 million bank guarantee for the term of the contract.

TSK Steel have reacted positively to the escrow proposals and provisionally accepts the principle of maintaining an escrow account in a Western bank, three months in advance of an agreed payment pattern. Payments into the account will be monitored and the bank will be under an obligation to report any default to the UK company. Payments to the UK partner will be made according to a previously agreed plan.

In order to facilitate the operation of this plan, the UK company also plans to help its Kazakh partner sell the steel through its contacts with a marketing organization based in London. This company has strong contacts in the former USSR, particularly with senior personnel at the Karaganda Steel complex.

The advantage of the escrow financial system for the UK company is that it will be informed by the bank if the account is in deficit. Karaganda's safeguard is that it only pays for the equipment with which it is supplied.

- Current status of the agreement: This agreement is in the process of being secured. The UK company still require certain 'comfort' documents to be signed by appropriate officials and feasibility studies to be approved. It was reasonably confident that agreement would be reached in the near future. Uncertainties have materialized, however, because of the speed of institutional changes in the former USSR. The UK company had spent much time negotiating with the Ministry of Metals in Moscow which was being run by an ex-managing director of Karaganda. A good relationship had been developed by the company, which it felt may have aided its cause in the face of any rival bids from the competition. Unfortunately, this ministry has now lost the status and power to influence affairs that it once held. The UK company was, however, impressed with the government of Kazakhstan and the way it had conducted negotiations.

  The company felt that the presence of an established Moscow office had been beneficial in easing the inherent administrative and practical problems of doing business with the former USSR. When extensive discussions were required, though, it was felt to be cheaper, easier and a more efficient use of time to invite CIS personnel over to stay in the UK rather than visit the former USSR.

2. Lipetsk: Situated 200-300 miles south of Moscow, the Lipetsk steel plant already utilizes continuous casting and earns foreign currency through its exporting. It impressed the company as it appeared to understand the changes which would be necessary for it to increase its share of the export market. It is eager to adopt more advanced technology and become more Westernized in its approach.

The discussions on the modernization of Lipetsk are less advanced than those concerning Karaganda. It is expected, however that the future agreement will be arranged on a similar basis albeit on a smaller scale.

The problems prevailing in the Soviet steel industry are also evident in Eastern Europe. In order to become competitive each plant needs to specialize and produce a certain type of steel to a high standard. It was felt that Eastern Europe was generally easier to conduct business with than the former USSR because of its more Westernized perspective. However, no concrete agreements have been made by the company in Eastern Europe, though many possibilities are being explored.

Company 2 - a steel rolling mill designer and manufacturer

The activities of this sister company include the design, supply, manufacture, installation and commissioning of all types of steel rolling mills throughout the world.

Commercial relationships with Eastern Europe and the USSR in the 1960s and 1970s

In the late 1960s this company sold an ore pelletizing plant to the USSR and another company within the corporation subsequently sold methanol plants in that market. Furthermore, this company gained a series of orders in the 1970s from the USSR for forging presses and manipulators involving both their design and build.

Commercial relationships with the former USSR and Eastern Europe in the 1980s and 1990s Commercial contacts have been maintained in this market-place, although the nature of the business has been changed as the UK company no longer produces presses. The company remains optimistic about the long term potential of the former Soviet and East European markets and over the years has put much effort into marketing itself in this region by visiting other steel works, giving seminars, making contacts and evaluating opportunities.

Since 1985 this company has received contracts from two Soviet steelplants, three contracts with a steelworks in Hungary and has carried out a feasibility study followed by a small follow-on contract with a Polish steelworks.

In 1988, this company was awarded a contract with the Magnitogorsk Iron and Steel Works Combine worth an estimated £13 million. The Soviet side, Tyazhpromexport,[6] had been liaising with the East German authorities concerning the provision of another steel mill in East Germany, and the Polish authorities regarding the construction of a new mill in Poland. The proposed East German plant would have far exceeded the production requirements of that country and hence it was expected that two thirds of the output of this new mill would be utilized by the USSR. The UK company was involved in discussions regarding the possible supply of equipment for this new plant in East Germany. However, after protracted and fraught discussions with the East German authorities, the Soviet side decided to cancel the East German and Polish projects and instead build the hot strip mill in the USSR at the Magnitogorsk steel works. Because of its contact with Tyazhpromexport on potential work with the East German steelworks, the UK company was then approached for certain components, which were not available in the USSR, for the Magnitogorsk project. This contract for the refinement of the 2,000mm. wide hot strip mill involved the provision of automatic gauge control for seven finishing stands; and the supply of an Encopanel heat retention system, two coil strapping machines and spindle lubrication. The company has completed and delivered the order, but as the mill has still to be erected at the time of writing, the company was awaiting instructions concerning installation. Payment for the delivery of the order has, however, been received from Tyazhpromexport. The company hopes that this contract at Magnitogorsk will enhance its profile in the former USSR thereby aiding its prospects for future contracts in that country. One contract has been gained since then for the supply of automatic gauge controls for one of the hot strip mills at the Karaganda steel works.

In Hungary, the company has been awarded three contracts with the Dunai Vasmu steelworks. In 1987, the company supplied a coil box unit in the hot strip mill and this was followed by orders for automatic gauge controls for one hot mill and one cold mill. As part of the original contract the company agreed to have groups of Hungarians visit the company for three months at a time to learn English and study UK management practices. At least three groups of Hungarians, with four in each group, took advantage of this opportunity and the company also employed a Hungarian electrical engineer. Hence, a solid relationship has been established with the Hungarian steelworks which the UK company hoped would enhance its prospects of gaining any future business.

The company also gained some work as a result of the Polish government's desire to modernize its main steelworks, the Huta Sendzimira works. Biprostal, the study and design office; Centrosatz, the office of foreign trade; and Hutmasproject, the design bureau, were all asked to formulate plans to modernize the steelworks. Their main proposal was for the works to invest in a large slab casting unit to improve efficiency and output and then sell the final product abroad to pay for the unit. However, no financial backing could be found to undertake this plan, as banks were not willing to risk investing the outlay of $60-100 million required to purchase the unit, on the basis of future sales of a Polish organization. The UK company was therefore commissioned to formulate a more realistic proposal. It subsequently advised a less dramatic approach of gradual modernization involving modest investments which could be financed at each stage. The company was paid for this study and was then asked by the steelworks to undertake another project concerning the type of design of loopers which should be adopted on the hot strip mill, the addition of loopers being the first of the company's progressive recommendations. The UK company provided the basic designs which were used on a similar mill in the UK and then supervised the modification of these designs which was undertaken by the steelworks.

Case study no. 2 - a foundry in the aerospace industry

Commercial relationships with the USSR and Eastern Europe in the 1960s and 1970s

The investment casting plant of an aerospace company purchased a Soviet process for investment casting in the late 1960s. The coating stage of the investment casting process consists of the application of a number of separate coats, followed by a sealing coat. The machine built under licence from the USSR contains 400 workstations which are loaded sequentially with pre-coated wax 'trees'. The workstations are attached to a continuous chain which passes several times through two coating stations and drying units, to build up a series of further coats on the wax trees. The process is completely continuous and automatic, apart from initial loading and unloading, and is capable of progressing through a complete cycle of production of 400 shells in seven hours, or 1,200 shells per day on a three shift basis.

Commercial relationships with the former USSR and Eastern Europe in the 1980s and 1990s

The plant was installed in 1970 and is still available for use today. It is currently not in operation because of a downturn in sales of spares, but can be used when an increase in production levels is required.

At present no specific machinery from the former USSR is being considered for purchase but efforts are being made by the company to be kept informed of current machinery and technology in that region. Hence, written and personal requests have been made to organizations in the former USSR for quotations on components and materials. Most of these companies have yet to respond, but one area where there may be co-operation, is that of coating technology. The UK company has sent turbine blades to the Central Institute of Aero Engines (TsIAM), who then coated them. The UK side is currently testing these coatings with regard to a possible future purchase.

Comments

Of the five companies described in this chapter, only one has continued to use the process originally purchased through the licensing agreement of the 1960s and 1970's, and that process is only used very rarely. The other four companies have concluded their licensing work, and three of these have neither maintained nor expanded their links with the former Soviet Union and Eastern Europe.

The exception to this general trend is the company described in case study 1, part of a large engineering group which has radically expanded its links in the region, mainly through the efforts of its recent acquisitions in the steel industry. These companies have been able to capitalize on the demand for steel in the former Soviet Union, the existing capacity which requires technological updating, the availability of existing natural resources of ores and other materials, and the potential for sale of steel in the West on a price effective basis. The development of these capabilities, as in the case of those successful companies described in chapter 3 above, has required a high level of investment in corporate resources, and the ability to develop a financial framework to support the business operation.

Notes

1. Hill, M.R., *East-West Trade, Industrial Co-operation and Technology Transfer*, Gower, Aldershot, 1983, pp. 133-147.
2. Cooper, J., 'Iron and Steel' in Amann, R., Cooper, J. and Davies, R.W. (eds) (1977) *The Technological Level of Soviet Industry*, Yale University Press, New Haven, Conn. and London, 1977, pp. 83-118.
3. *Ibid.*
4. *Ibid.*
5. Aylen, J., (1990) 'Neglected Steel - The Future of Steel in Eastern Europe and Developing Countries', 1990 p. 11, (unpublished report).
6. The Soviet foreign trade organization specializing in the export of heavy engineering equipment.

6 Technical co-operation

Introduction

This chapter describes the activities of five companies which have been engaged in technical co-operation agreements related to joint research and development, and the exchange of technical information, with industrial and scientific organizations within the former socialist countries of Eastern Europe.

The activities of these companies were originally studied to provide further information on the westward flow of Soviet and East European technology at the company level.[1] The sample of companies was as follows:

- an electrical engineering company;
- a photographic equipment company;
- the National Coal Board;
- an oil company;
- and an electronics company.

The electronics company is a manufacturer of electronic components and finished products, which has had extensive experience of trading with the socialist countries of Eastern Europe with sales of licences being made to Bulgaria, GDR, Poland, Romania and the USSR. It signed an agreement for technical co-operation with the Soviet State Committee for Science and Technology in 1968, with the main technical areas to be covered by the agreement being quite broadly defined, relating mainly to communications and electronic components. The company has recently undergone major internal change both in terms of ownership and product range which has altered its trading relationship with the Eastern bloc dramatically. It now solely provides military products and was therefore not allowed to conduct business with either the former Soviet Union or Eastern Europe at the time of writing, although it does not discount the possibility of conducting business with the region in the future.

Case study no. 1 - an electrical engineering company

Commercial relationships with the USSR and Eastern Europe in the 1960s and 1970s

This large electrical engineering company signed an agreement in 1975 with the Soviet State Committee for Science and Technology, covering the exchange of scientific and technical information for three main technologies - power engineering, communications and electronics. Under the terms of the agreement, the company's communications division engaged in discussions with the Soviet State Committee for Television and Radio (Gostelradio) and the Ministry for Communication Equipment, on new types of colour television cameras and charge couple devices. During these discussions particular attention was focused on Soviet requirements for the then forthcoming Moscow Olympics. Although the company was disappointed by the overall lack of commercial agreements emanating from the discussions, the electronics division did obtain a contract for colour monitoring equipment to be used in the 1980 Olympic Games. This technical co-operation agreement was supplemented in 1976 and renewed in 1981 but there is now no record of its current operation.

Commercial relationships with the former USSR and Eastern Europe in the 1980s and 1990s

This firm has formed many business oriented agreements with the Soviet Union and Eastern Europe in recent years, although these are unrelated to the aforementioned agreement. The most active Soviet project is a joint venture between the UK firm and the Moscow telephone network, set up to operate international pay-phones through telephone-cards sold for hard currency. These cards can only be used in the telephones provided by the UK partner which have been installed in appropriate points by the Moscow telephone network.

The idea for this joint venture came from the British company who approached the Soviet side. Such a development would have been impossible a few years ago when foreign participation in the Soviet telecommunications sector was strictly forbidden. However, international political initiatives following Mikhail Gorbachev's election to the post of General Secretary of the Soviet Communist Party and a willingness by the British government to respond to these initiatives through the support of trade and technology transfer, opened up the possibilities for collaboration in this sphere. It took some two years of discussion and negotiation before the joint venture company was established in 1988.

The UK company has supplied one hundred card-phones working on the magnetic cards which are unique to the software installed in these telephones. To date, seventy have been installed and the most favourable locations for the other thirty are being evaluated. The UK partner also provided training as part of the agreement and sent three engineers to Moscow to install the telephones, to ensure the computer software was functioning and to train Soviet personnel in product maintenance procedures. One UK engineer is still working in the former USSR with four CIS engineers under his supervision, and CIS employees of the joint venture company continue to be sent periodically to the UK for training in legal, financial and business matters. For its part the Moscow partner found suitable sites in key business and tourist locations to locate the telephones, provided the labour to connect them to the international network and organized the initial selling points for the cards.

The joint venture company in Moscow employs four resident British personnel and ninety-two CIS personnel, and is organized in departments like most Western companies. As well as organizing the sale of the metallic cards it also markets digital switches and sophisticated telephone equipment. Currently these products are bought by the company from the UK partner but there is no obligation to continue this trend. The joint venture company also offers a 24 hour repair and maintenance service for their telephones and fax machines.

The joint venture's main business at present involves the sale of the telephone cards, mainly to Western businessmen and tourists for hard currency, and this part of the business is currently operating on a highly profitable basis. The revenue has increased steadily from month to month since its inception eighteen months ago with no immediate signs of reaching saturation point. Following the decision to adapt the telephones to accept credit cards, agreements were made with each of the largest credit card firms from January 1991 making the telephones even more accessible to Western personnel and ensuring that continued profitability was maintained.

The British partner mainly sees this agreement as a platform for future business. Although the joint venture company is making a profit in financial terms, the management costs entailed in securing the agreement and the training of Soviet personnel have to some extent outweighed this financial gain. Accordingly, plans are underway to enable the UK company to take further advantage of the evident opportunities in the telecommunications market-place. These plans stem from a recognition that most of the problems and faults detected in the telephones are caused by faults in the antiquated Moscow telephone network which is struggling to cope with the current volume of traffic. The joint venture company therefore plans to set up a satellite link to the UK company's own exchange using fibre-optic cable to ensure the Moscow telephone network is by-passed completely. It is hoped that use of this link could then be sold to British companies working in the former USSR, as the inadequate communications infrastructure is one of the key criticisms propounded by British companies currently doing business in that region. This service would aim to offer a telecommunications link in the former USSR to Western standards. In the first year it would operate with 1000 lines but this number would be increased to 10,000 lines by the end of the year, if all goes according to plan. Current discussions are focusing on whether the joint venture company or the UK company will provide the necessary finance to initiate work. The technology will be provided by the latter while the former will operate and maintain the system once it is in operation.

The British partner felt that the initial joint venture agreement appealed to the Moscow telephone network for a variety of reasons. In the first place it provided up-to-date working technology in the form of modern quality card-phones with international connections, which was not previously available in the USSR. The international credibility of the telephones also ensured that Westerners would be likely to use them and therefore would purchase the cards, enabling the venture to operate on a profitable basis and provide much needed hard currency reserves. Furthermore, the agreement ensured that Moscow personnel would be trained by Western engineers in up-to-date technology.

The success of this Moscow joint venture has already been repeated in Eastern Europe. An almost exact replica, which became effective in May 1991, has been set up in Bulgaria with the Bulgarian Committee of Posts, Telecommunications and Information (CPTI). The initial five year agreement involves the UK partner supplying two hundred telephones to be connected in key tourist or business areas throughout Bulgaria - Sofia, the Balkan ski resorts and the Black Sea resorts. This agreement differs from the Moscow one only in that it covers the whole of the country.

In August 1991 trial card-phones were installed by the UK company in Warsaw with the hope of establishing a joint venture there in due course. Approaches have also been made to Hungary and Czechoslovakia but with little apparent success as organizations in these countries have been buying pay-phones themselves, and are not interested in co-operation with Western companies in this area.

While these two joint ventures were considered by the UK company to be successful it was emphasized that there were many other potential projects in the former USSR, which were not going ahead. The main problem was a lack of finance available to the CIS buyer, but this was compounded by the difficulty of not knowing who to contact in order to remedy the situation. For example, another offshoot of this UK company signed a licensing agreement for an electronic component. The licensee, was acting on behalf of a ministry who represented the CIS factory wishing to purchase the licence. While the agreement has been signed, the technology has yet to be supplied because payment has not been received. The organizations in the former USSR are adamant that they have deposited the requisite roubles but the bank will not confirm this. With four parties involved in the agreement it is difficult to define a solution to the problem. However, this company remains confident about the overall future prospects for trade with the former Soviet Union and Eastern Europe. It is prepared to invest in the short term in the belief that it is gaining a foothold in a vast market-place devoid of modern technology.

Case study no. 2 - a photographic equipment company

Commercial relationships with the USSR and Eastern Europe in the 1960s and 1970s

This company, as a designer and manufacturer of high speed cameras and associated photographic equipment, signed an agreement with the Lebedev Institute of the Academy of Sciences of the USSR in 1977. This agreement was for co-operation in the field of research and development of ultra-high speed cameras of the image converter type. These cameras, which are capable of diagnosing events in the pico-second range, are used mainly for laser research directed towards the production of electricity from fusion. The USSR had marked superiority in the production of image tubes, whereas the UK was stronger in the design of the required electronic driving circuits.

The agreement emerged from a number of technical and business contacts between the company and its Soviet partners over the previous twenty years, and provided for the exchange of scientific and technical information, the co-ordination of research and development, and the training of personnel. It was also intended to engage in related trade in licences and components as the Academy Institute was interested in the purchase of some of the company's electronic drive units/camera bodies, and the company was interested in the purchase of Soviet-developed specialized infra-red cathode tubes.

This intention to engage in related trade in licences and components had encountered certain problems, however, because of factors which prevented a straightforward import/export transaction, as the tubes in question were manufactured by a laboratory responsible to the Soviet State Committee of Standards and not the Lebedev Institute. The Institute was therefore unable to export the tubes through an appropriate foreign trade organization to generate sufficient foreign currency for the purchase of the required number of electronic drive units, as it was not the manufacturer of the components.

Commercial relationships with the former USSR and Eastern Europe in the 1980s and 1990s

The commercial arrangement negotiated in 1987-88 to overcome these problems, involved the exchange of drive units for cathode tubes through a barter type arrangement. The agreements between the company and the General Physics Institute (IOFAN) of the Academy of Sciences (formerly the Lebedev Institute) were arranged under the umbrella of the 1977 agreement and were set up by the 'Licensintorg' foreign trading company,[2] which also acted as the barter house. As the cathode tubes are very sensitive and need to be hand carried, representatives from the UK company have made regular visits to the USSR to collect them.

It is still the case that there is no design available in the UK to match the particular spectoral response of the Russian cathode ray tubes, and the company has consequently managed to continue the agreement, despite changes in company ownership and product range.

Company background In 1983 the owner and managing director of the company retired, resulting in a management buy out by the company's executives. Since that date the name of the company has been shortened and its product offering has increased to include airborne military videos, and a graphic arts service and sales for pre-press equipment.

Development of the 1977 agreement The tenets of the original agreement still apply and therefore there has been continued exchange of scientific and technical information, the co-ordination of research and development and the training of personnel. Executives of the UK company have also lectured in the Academy Institute on a variety of topics including capitalism, responsibility and accountability.

The changes in the former USSR since Mikhail Gorbachev's election to the post of General Secretary of the Communist Party in 1985 have influenced the agreement in certain ways however, chiefly as a consequence of the uncertainty regarding legislation and the changing roles of the main bodies. There is now no legal need for the intermediary contribution of Licensintorg as individual Russian establishments have become more autonomous; although it was considered by the company that the laws have been changing so rapidly in the last 6-8 months that very few Russian officials are completely clear on the status of the current legislation regarding foreign business ventures. Laws were apparently passed in early 1991 making barter type deals illegal but the UK company, judging from its own experience, felt that these were not being enforced.

It remained the case that negotiations between the two parties were overly lengthy because of inadequate business communication systems - correspondence is conducted by telex as the Russian fax network does not yet operate efficiently. The UK executive also felt that the process of securing an agreement was frustratingly long and drawn out because of the Russian partner's insistence on writing agreements for each decision made on co-operation.

However, the UK partner felt that it had gained the trust and respect of the Russian partner through the many years of co-operation and this was being rewarded in other ways. For example, the Academy Institute set up a new machine shop inside the Academy and decided to purchase its own supplies. Over the past 2-3 years, whenever the Institute has had hard currency and required Western goods, it has purchased goods through the company. So far the orders have exceeded £1/2 million and have involved a whole variety of low tech, high quality goods including floppy disks, personal computers and drilling machines.

The UK partner did not foresee any expansion in its future dealings with the former USSR for a variety of reasons, including the fact that the majority of its products are still ineligible for trade with the former USSR and Eastern Europe because of Cocom[3] regulations which continue to apply. These regulations ensure it can only deal on a low-tech basis. Even if these controls were lifted in the future, however, organizations in the former USSR and Eastern Europe would not be in a strong position to negotiate a deal. Although their demand for Western products and expertise was insatiable they have neither the money, products or technology to attract Western companies.

Case study no. 3 - British Coal technical services and research executive

Commercial relationships with the USSR and Eastern Europe in the 1960s and 1970s

As the Mining Research and Development Establishment (MRDE), this organization signed agreements for technical co-operation with the ministries responsible for the coal industry in Poland (1974), the USSR (1975) and Czechoslovakia (1979). This case study charts the subsequent development of these agreements.

Technical co-operation with the Polish coal industry (1974) This agreement was originally signed as a consequence of Polish expertise in the construction of new mines, including the design of standard modules for that purpose. The only co-operation carried out was an exchange of instruments, and currently this agreement exists in name only.

The Polish side remained interested in technical co-operation with the UK side and proposed to supply equipment and personnel for the establishment of new pits. However, the UK party was sceptical about the commercial viability of such proposals and they were not adopted.

Technical co-operation with the Soviet Ministry for the Coal Industry (1975) The Soviet agreement involved a series of working groups covering tunnelling machinery, reliability testing, dust control, computer applications and software systems, meeting once or twice a year to decide on the form and scope of co-operation. The majority of the co-operation took place in the areas of reliability testing and tunnelling machinery.

The collaboration on reliability testing grew from initial discussions in the Anglo-Soviet Working Group on the Reliability of Mining Machinery. These revealed that the State Project Institute for Mining Machinery in Moscow had developed advanced techniques for reliability measurement while the MRDE had better testing facilities and practical experience in the evaluation of these techniques. It was decided, therefore, to test out these techniques on coal shearer gearheads, which are critical assemblies influencing the reliability of mining machines. Both parties carried out tests, the results of which were then exchanged. This information was beneficial to both sides, particularly as it was apparent that testing times could be halved for these products.

The co-operation in tunnelling machinery led to technical collaboration in the development and joint prototype production of a particular type of road heading machine. At the time of drawing up the specification, in 1975, it was intended that the machine would be an advanced model, particularly in terms of power. It was decided that the Soviet side would design and build two sets of the feeder unit while the British side would design and build two sets of the cutter unit, with subsequent exchange of one set of both units. Unfortunately, due to a variety of factors which prolonged progress on co-operation and the simultaneous development of more advanced machinery by British equipment manufacturers, the machine was technologically outdated by the time it was completed.

Technical co-operation with the Czechoslovakian coal industry (1979) The agreement with the Czechoslovakian coal industry involved collaboration in the following areas:

- rock testing and behaviour in heading machines;
- general problems in testing and the particular problems of testing powered supports under cyclic loads;
- seismic ranging, with particular reference to detection of faults ahead of the seam.

In general, the operation of this agreement appears to have been beneficial to both parties. There was a useful exchange of information on rock cutting with a Czechoslovakian authority, and some Czechoslovakian rock cutting picks were tested. In the case of powered shields, the Czechoslovakian partner had more experience in the use of shield supports which was beneficial to the British side who had the better testing facilities. The work on seismic ranging encountered certain problems, however, as a result of the differences in techniques for signal analysis used by both sides and the existence of a series of patents held by both parties.

The agreement was operated on an annual exchange visit basis to pits and equipment manufacturers, with the hosts covering the residential costs of the guests and the provision of a spending allowance. The main criticisms expressed by the British side regarding the operation of this agreement, concerned the slowness of the Czech bureaucratic system and the difficulty of not being able to contact individual scientists quickly by letter or telephone.

Commercial relationships with the former USSR and Eastern Europe in the 1980s and 1990s

Company background This organization has undergone profound changes in its mode of operation. From a research body of a nationalized industry with a responsibility to transfer technology as required, it now operates as a business as Technical Services and Research Executive (TSRE). It competes against external organizations to sell its technical services to British Coal and others in the UK and abroad.

This change has clearly influenced its approach to technical co-operation. Whereas before it entered collaboration agreements to share expertise, now it must evaluate the commercial viability behind such collaboration. This change in outlook has significantly affected its contacts with the former Soviet Union and Eastern Europe.

Current commercial relations with Poland Through the UK 'Know How Fund' British Coal organized a series of management training courses for the Polish Coal Industry from February 1989 to February 1990. It ran eleven courses in total - three in the UK and eight in Poland - with thirty-six people on each course including senior management in Polish mines and senior officials in the Polish hard coal industry. These courses were felt to be beneficial to the Poles who were interested in the experience of the British coal industry since the end of the miners' strike in March 1985. Particular attention was paid to the management of the restructuring programme that had taken place, and the ways in which manpower had been reduced, including the assistance given by British Coal in finding alternative employment for redundant employees. The 'Know How Fund' ensured that British Coal covered its costs but when no further funding was available from that source, the courses stopped.

Current commercial relations with the USSR British Coal was sceptical about the overall results of the working groups set up under the 1975 agreement and felt that this Soviet agreement would not have been signed by British Coal in its present form. It was deemed to be a very one-sided agreement with no tangible benefits to the UK side because Soviet technology was so out-dated. In many areas the coal industry in the former USSR currently lags behind the UK in respect to the level of technology operated in its coal mines.

This agreement has now been replaced, though not on the initiative of British Coal. A large scale agreement to exchange technology on energy related matters was signed by the then Secretary of State for Energy, (Peter Walker) with his Soviet counterpart in 1988. A sub-agreement was to cover coal with a steering group including representatives from British Coal and mining equipment companies, ABMEC[4] and DTI. After a visit to the USSR and a return visit by the Soviet delegation, five working sub-groups were established, to cover the following five areas for co-operation:

- coal preparation;
- the use of computers in mining;
- transport (non-British Coal);
- reliability;
- coal-face (non-British Coal).

The success of collaboration in the above areas has been mixed. The sub-groups were to meet their Soviet counterparts every eighteen months to agree a programme of work to carry out, but in practice this has often been difficult to achieve.

In the coal preparation agreement there has been some exchange of technical information with TSRE agreeing to test an instrument the Soviet side had developed. TSRE was not interested in its purchase, however, as it was not considered to be a sufficiently advanced model. There has also been some exchange of technical information in the agreement relating to the use of computers in mining. However, the agreement is currently at a standstill. The UK side was due to visit the former USSR two and a half years ago and is still ready to fulfil this commitment, but the Soviet side has consistently delayed the proposed visit. The reliability agreement involved an exchange of information and a programme of work. This agreement is also at an impasse at the time of writing. The Soviet delegation was expected in the UK some two years ago but the visit failed to materialize.

The control of the overall agreement has been transferred to the DTI who are eager for the steering group to meet again. TSRE was querulous about the chances of this producing positive results however, because it could not overcome what it believed to be the main problem in collaboration - namely the insufficient technological know-how of the former USSR. It felt that now it was operating on a commercial level as TSRE, the government would be less able to draw it into an unprofitable obligation.

Current commercial relations with Czechoslovakia The 1979 technical co-operation agreement has not been terminated but there has been no co-operation since 1984. The collaboration in the areas listed above was exhausted and although the Czech side suggested areas for future co-ordination at the time, they were not taken up by the UK side. The timing was not favourable for extending the agreement - it was 3-4 weeks before the commencement of the miners strike of that year - but in addition the areas suggested were of little interest to the UK side.

The possibility for future co-operation with the former USSR and Eastern Europe was not completely discounted by TSRE. It receives many delegations from the former USSR interested in co-operation so is well in touch with the needs and resources of the coal industry in that country. It was felt that other opportunities for co-ordination may arise out of the 'Know How Fund'.

It did stress though, that the agreements signed in the past were luxuries which it could not afford today. It did not feel that many profitable ventures could be achieved with either the former USSR or Eastern Europe for many years. Although the individual mines in the former USSR now had more autonomy and could sign their own agreements with the West, they were unlikely, it was felt, to possess sufficient hard currency and would want to trade technology for their coal.

TSRE felt opportunities in Eastern Europe (particularly Poland, Czechoslovakia and Hungary) were more promising than those in the former USSR as the technological levels of their mines were ahead of their former Soviet counterparts. Again, it was emphasized that it would take many more years before these countries understood Western business practices and applied them to their mines in terms of manpower requirements, productivity levels and a pricing policy based on international standards. Until this happened the possibility for profitable commercial agreements with these countries was considered to be unlikely.

Case study no. 4 - an oil company

Commercial relationships with the USSR and Eastern Europe in the 1960s and 1970s

This company is a large multinational corporation engaged in the oil and chemicals business. It signed a five year technical and scientific co-operation agreement with the Soviet State Committee for Science and Technology, in the 1970s.

This umbrella agreement was appended by three subsidiary agreements with the Ministry of the Refining and Petrochemical Industries, the Ministry of the Oil Industry and the Chief Administration of the Microbiological Industry. The first of these agreements relating to refining was concerned with the exchange of information on general operating practices and trials related to certain processes. The company found the discussions on refining operating practice to be generally useful, and the examination of processes helped to familiarize each side with the other's capabilities and highlighted interesting differences in feedstocks and specifications.

The second agreement, related to oil field operations, particularly the general problem of recovery, especially from reserves with heavy oils. Again, some interesting information was transferred on thermal recovery methods.

The final agreement concerned the manufacture of synthetic proteins using technology which the company had installed in a Western European plant. A useful exchange of technology materialized, but the anticipated sale of technology to the Soviet side for the construction of a synthetic protein plant did not come to fruition.

Commercial relationships with the former USSR and Eastern Europe in the 1980s and 1990s

The co-operation in the above three areas has not continued, but the company remains actively involved in the former USSR and East Europe.

The company is currently considering the potential for oil production in Western Siberia through the development of existing oil fields and the exploration of new sources. It was still not clear, however, whether the amount of oil to be found in Eastern Siberia would justify the cost and resources involved in exploring oil fields in that region.

Representative offices have been established in both Alma Ata in Kazakhstan and in Moscow, staffed with both British and local personnel. It is hoped that they eventually will sell the company's lubricants. The company also runs a small chemicals office in Moscow for the sale of imported chemicals. The company has also been involved for many years in a consultancy agreement covering poultry breeding, with the district council in Stavropol in South Russia.

In August 1991, the company became involved with the USSR in a Soviet-British project involving another UK company, a major engineering contractor (see case study no.3, chapter 3). Under the contract, the Soviet fabrication yards, Shelfproyektstroi, based in Baku, were to supply templates for oil rigs in the UK, which would be used by the UK company in the newly developed North Sea oil fields. The deal provided the Soviet side with a rare export opportunity in a manufactured product and the potential of future business, as the long term aim is to fabricate a complete platform fabricated in the former USSR.

The company has also invested in education and training in the former USSR and is accredited by its former partner, the State Committee of Science and Technology, now part of the Ministry of Higher Education. It has provided two fellowships for Russians to study at the London Business

School, and has also conducted seminars for the oil industry in the former USSR on a wide range of topics including marketing and refining.

The company has traded with Eastern Europe for many years now and has representative offices in both Hungary and Czechoslovakia.

Comments

Of the five companies previously involved in technical cooperation with the former socialist countries of Eastern Europe, only one company (case study no. 2) is still involved in technical co-operation with the same partner. Internal change within the company was the main reason for the decline of technical cooperation in two cases. In case study no.3 there was a need for agreements to be profitable, as the organization has moved from a position of providing a service to a nationalized industry to one of being more cost effective. The electronics company is presently prevented from trade in the Eastern bloc because its product range is mainly military in application. This position could change, however, with the major changes in relationships taking place between NATO and the former members of the Warsaw Treaty Organization.

The other three companies are continuing to invest in the market-place and may feel they have benefited from their past association with the region. In the case of the company described in case study no.1, its previous contacts provided a means for establishing a novel type of joint venture with future business potential which has been gradually built on and extended to good effect.

Notes

1. Hill, M.R., *East-West Trade, Industrial Co-operation and Technology Transfer,* Gower, Aldershot, 1983, pp. 148-59.
2. The Soviet foreign organization specializing in the international purchase and sale of industrial patents and licences.
3. Coordinating Committee for Multilateral Export Controls. The membership of Cocom comprises all members of NATO, with the exception of Iceland but plus Japan, and more recently Australia. The major objective of Cocom has been the denial of militarily useful commodities to the Warsaw Pact countries and China. (see Hill, M.R., *Soviet Advanced Manufacturing Technology and Western Export Controls*, Avebury, 1993, pp.7-12.
4. Association of British Mining Equipment Companies.

7 East European companies in UK

Introduction

This chapter provides a descriptive account of the commercial activities of four British companies which are either partly or fully owned by foreign trade organizations of the former socialist countries of Eastern Europe. These case studies are follow-up accounts of four of five companies previously researched by one of the present authors in 1986.[1] It proved impossible to arrange a meeting with the remaining company from the original sample, a distributor of Polish shoes and leather products.

The previously published material documented the management policies and practices of those five British companies as part of an overall study to assess whether enterprises from the socialist countries of Eastern Europe and the USSR could be properly termed 'Red Multinationals', in view of their business activities in the West.

This company-level research was intended to supplement information previously compiled by McMillan[2] on an aggregated and global basis. The original case studies were compiled after interviews had been conducted with senior executives in late 1982 and early 1983. These follow-up case studies, therefore, describe the development of the companies following a time span of ten years, in which the structure of foreign trade organizations in Eastern Europe has undergone a profound change, as described previously. These case studies present an account of the effect of this change from the perspective of companies operating in the UK having the ties of ownership with Eastern European foreign trade organizations.

Case study no.1 - The London Chemical Company Ltd

Company Background

This company was established in 1972 as a joint Hungarian-British company for the sale of chemical products. The original ownership was 50 per cent Hungarian (Chemolimpex foreign trade organization) and 50 per cent British (Tar Residuals Ltd.), but in 1979 it was changed to 80 per cent Hungarian and 20 per cent British (renamed TR International), with an increase in capital. In September 1990, when the British company TR International was bought out by a German company having connections with the Polish chemical industry, the company became fully owned by the Hungarian partner - Chemol RT, the adoption of a new name reflecting the changed nature of the former Chemolimpex foreign trade organization. Chemol RT is now a joint stock company and is due to be privatized in the near future.

The London Chemical Company currently employs seventeen personnel including three Hungarians, two of whom hold senior positions. This compares to the twelve people employed some ten years ago.

Company activities

While still primarily engaged in the sale of chemical products in Hungary and the UK, the company's business activities have changed considerably during the last ten years. UK exports to Hungary continue to account for £5 million worth of the company's overall business, but the proportion of Hungarian products that the company imports to the UK has increased dramatically from £4 million of the company's overall turnover, to the current level of £20 million. The company continues to sell mainly to the UK and Irish markets to end-users and occasionally to distributors, but the proportion of sales to third markets is negligible. The company has become more involved in distribution, as a result of its experience with Hungarian-made tyres through which it progressed from being a simple importer of tyres to a stockist able to provide a complete after-sales service (including claims and warranty) and capable of opening new accounts. It has marketed polyethylene and polypropylene in the UK since 1988, and PVC since 1978.

Another primary change experienced by the company during the last few years was the transition from operating on a commission basis, to using its own account in each transaction. Previously, exports to Hungary were handled on the company's own account for UK firms, as were one third of the Hungarian exports, but the remaining two thirds were dealt with on a commission basis. The company felt that this change was evidence that it was a profoundly different type of organization compared to ten years ago. Whereas before the company's primary objective was to provide a service to the Hungarian government and to facilitate its purchase of requisite raw materials through the sale of Hungarian goods for hard currency, the company's main purpose is now to generate profit through its transactions. It is now felt, therefore, that although the commercial aspect of the business has been present for years, it is fundamental today, with the company now operating in the same way as any British trading company.

The company also felt that its stature had altered as a result of the reduced role of foreign trade organizations in Hungary, and the corresponding eligibility of Hungarian producers to sell directly to their preferred customers. Previously, the company acted as sole agents for almost all Chemolimpex traded items, and also acted as sole agents for other foreign trade organizations which sold products which had previously been sold by Chemolimpex. Chemol RT is no longer the sole agent for the Hungarian chemical industry, however, and having experienced contraction in terms of size, turnover and scope of its activities, is now mainly engaged in the import of products into Hungary. Despite the fact that Chemol RT owns the London Chemical Company, its changed status has not affected the company's business, and 95 per cent of all Hungarian chemical exports to the UK continue to be conducted through the London Chemical Company. This illustrates that the company is more autonomous than before, has established a reliable reputation for itself, and possesses greater financial strength.

The company's profit levels have grown from £20,000 around ten years ago, to £100,000 currently. Although it has sometimes been troubled by supply difficulties in sourcing its products from Hungary, the company's main preoccupation recently has been finding customers to purchase its products during a period when both the UK recession and the decline of the chemical industry have been affecting sales. It was felt that the depression in the chemical industry had created a situation of over-supply in the business which had consequently led to suppliers undercutting prices. This factor, it was felt, negated the cost advantages the company was perceived to possess by importing from Hungary. Moreover, the cessation of cheap raw materials supply to Hungary from the former Soviet Union meant that Hungarian supplies were more expensive now than in the past. The company, therefore, had to adapt to these conditions and began to import products from other countries and to sell into new markets. As a result of a considerable expansion of this trade, the company established a separate company in January 1992 - 'Albion Chemical Trading Ltd' - specifically to undertake this third country business.

The company, in conjunction with Chemol RT, also set up a company in Czechoslovakia in 1992 to administer Czech and Slovak chemical imports to Hungary. Although that trade had always existed, the office, staffed with personnel previously representing Chemolimpex in the Hungarian market, was felt to facilitate the establishment of more personal relations with Czech producers. This had been unnecessary in the previous system whereby trade could be conducted through the appropriate foreign trade organizations.

The company's role as the intermediary for UK manufacturers exporting to Hungary has also evolved over the last ten years, partially in response to the resistance it encountered from some of its British clients engaged in export to Hungary. They considered that the cost of the company's services made their products less viable in the extremely competitive Hungarian market. In response to those arguments, the company set up an office in Budapest in 1988 staffed with four of the company's personnel to provide a more effective service for its clients in Hungary. This office has also enabled the company to promote its services and to identify further business opportunities in Hungary.

The company felt that the advantages of a British base were as relevant today as before. Although it was felt that some of the suspicion and antipathy to Eastern European countries prevailing during the era of the Cold War had eroded since the late 1980s, the company also considered that it was unlikely that many Western businesses would want to conduct business with a company based in Hungary, for a variety of reasons including communication difficulties.

The company could not predict the likely effects of the proposed privatization of Chemol RT on its future business operations, or of the effects of further change in Eastern Europe. It believed that it had managed to adapt well to the changes affecting the former Eastern bloc to date and had improved its profit margins despite adverse conditions prevailing in the UK in general and the chemical industry in particular. It also felt that it now had an established name and reputation in the market which would inspire loyalty from both its suppliers in the UK and its customers in the UK, and also enable it to attract new business.

Case study no.2 - Toolmex Polmach Ltd

Company background

Toolmex Corporation UK was purchased by the Polish Metalexport and PZL foreign trade organizations in 1978. In 1981, the company purchased TI Polmach Ltd. which had previously been established at Coventry in 1976 as a joint venture between Metalexport, Daltrade and a subsidiary of Tube Investments Ltd, the latter being the major shareholder. Following this purchase, the company was renamed Toolmex Polmach Ltd, with Metalexport owning 90 per cent of the share capital and PZL, the remaining 10 per cent. The number of shareholders in the company has since increased with the addition of the Polish foreign trade organizations, Electrim and Intervis, in 1986. Metalexport now owns 73 per cent of the company's share capital, Electrim owns 12 per cent, PZL 8 per cent, and Intervis 7 per cent. The company employs forty-six people in total, thirteen of whom are Polish. This compares to the thirty-two the company employed in the early 1980s.

The company is currently divided into five divisions, with the addition during the last ten years of a forklift truck division and an export division to the established machine tools division, a small tools division and an industrial equipment division. Metalexport is still the principal supplier of the machine tools division, both Metalexport and Intervis supply the small tools division, PZL supplies the industrial equipment division, and the forklift trucks are assembled by the Polish company, FUT Suchedniou, using components purchased by the company in the UK. The formation of the forklift truck division in January 1990 increased the number of divisions in the company at that time to six, following the addition of a copper cable division in the mid-1980s. This division, supplied by Electrim, proved to be particularly profitable, however, and it expanded to such an extent that in August 1990 the company decided to convert it into a separate company - 'Copper Cable Company Ltd.', in which Toolmex Polmach Ltd. now hold a 20 per cent shareholding.

Company activities

The company's five divisions continue to be involved in the following types of business activity:

- the import of Polish products into the UK;
- the export from the UK of both Polish and British products;
- when possible, tendering for project planning in third countries, sometimes on a joint venture or industrial co-operation basis;
- after-sales servicing.

The company also carries out the normal insurance, freight and shipping activities associated with its role as an importer and exporter of industrial equipment. It still does not undertake any manufacturing activities in the UK, however.

The operation of the company's three established divisions has changed by varying degrees during the last ten years and their current activities are described below:

Machine tool division This division is still engaged in the import of Polish-built machine tools into the UK, for sale in that market and other markets using the UK as a base for re-exporting to third countries. However, the division's business has been adversely affected by the UK recession which has made potential customers reluctant to purchase large capital goods. The recession was also cited as the reason why the company's attempt to establish itself as a merchant for certain makes of British and German machine tools to complement its existing product range, failed to materialize.

The other major activity carried out by this division is servicing. The company provides guarantee services for new machines as well as the normal after-sales service facilities.

Small tool division This division is still primarily engaged in the import of Polish-produced small tools into the UK. Its product range remains similar to ten years ago and therefore includes precision measuring tools, hand tools, cutters and vices, carbide tools, and machine tool accessories. The vast majority of these tools are sold in the UK market to engineering companies and DIY stores, but on occasion some are re-exported to third markets such as the USA and France. The company is considering the possibility of sourcing supplies for this division from Taiwan in the future. This division has continued to operate successfully and its business has grown despite poor economic conditions prevailing in the UK.

Industrial equipment division This division was previously divided into two subdivisions, one concerned with contracts and consultancy, and the other with the sale of Polish diesel engines, generators and aircraft equipment.

Industrial equipment division: contracts, special projects and consultancy sub-division This is now a division in its own right - the export division -

and it continues to provide some of the services previously undertaken by the contracts, special projects and consultancy sub-division. Primarily, it has continued to act as the representative for a number of British companies engaged in the export to Poland of machine tools and associated production equipment, because of its well established contacts in Poland and the continued presence of the company's office in Warsaw. Although this business has reduced recently, the company has acted as the intermediary in concluding many important deals between UK and Polish companies during the last few years. The nature of this business has been affected by the recent reduced role of Polish foreign trade organizations in the conduct of foreign trade, as on occasion the company will deal with the Polish manufacturer directly rather than through the appropriate foreign trade organization. However, it was emphasized that many of the Polish foreign trade organizations, including Metalexport, continue to play a prominent and influential role in the conduct of foreign trade by virtue of their experience and contacts; and that they have endeavoured to become more professional and Westernized in their approach to business, which has enabled them to maintain their influence for the foreseeable future.

Unlike the previous contracts, special projects and consultancy sub-division, the new export division no longer acts for Polservice on an agency basis, and is not involved in the sale of licences. It has also failed to initiate any new industrial co-operation activities comparable to the two agreements signed in the early 1980s, in which the companies involved attempted to benefit from the cheaper labour costs associated with the manufacture of products in Poland. In both instances, the agreement was signed between the Western firm and Metalexport, with the contract and consultancy subdivision of the company acting as the Western firms' representative. In the case of the former agreement, the company was involved with Liftec S.A. of Switzerland, which was the holding company for Lansing Bagnall, a British manufacturer of forklift trucks. Lansing Bagnall supplied the design documentation and certain components, and the Polish partner produced other labour intensive components, assembled the final product, and provided an after sales service in Poland through its Warsaw office. Lansing Bagnall bought the finished trucks from its Polish partner which were then distributed through its own distribution network. In the case of the latter agreement with M and T Chemicals, a Birmingham-based manufacturer of electroplating equipment, the aim was to manufacture filters and centrifuges in Poland using both British and Polish components, which the chemical company would import for use in its electroplating equipment. However, this agreement did not progress as expected. This was blamed on the reduction in demand for this company's products, induced by the recession in the UK.

No other opportunities for similar industrial co-operation agreements have arisen in the last few years, although the company did not discount the possibility of concluding others in the future.

Industrial equipment division: diesel engines and aviation products subdivision This subdivision of the company is still engaged in the import of Polish diesel engines, power generators, the PZL range of gliders and light aircraft, and electrical vehicles for factory materials handling and golf buggies into the UK. It also continues to provide an after-sales service for its product range. This subdivision is currently experiencing a contraction in sales which was expected to be rectified when consumer confidence in the economy returned.

The company felt that the changes in Poland in the late 1980s and early 1990s had not altered its operation to any significant extent, although it considered that it had gradually adopted a more professional approach to business and now operated in a manner more reminiscent of a British company than a Polish one. The company felt that although the profit margins of some of its divisions had been eroded as a result of the effects of the UK recession, it was still operating on a profitable basis. It cited the continued profitability of the small tools division, the inauguration of the fork lift division and the establishment of the 'Copper Cable Company Ltd' as evidence of its continued success.

Case study no.3 - FLT and Metals Ltd

Company background

This company was established by its owners Impexmetal (83 per cent shareholding), Anglo-Dal (16.7 per cent shareholding) and Dal (0.3 per cent shareholding) in 1974; and its ownership and the distribution of its shares remain the same today, nearly twenty years later. The company also continues to employ some twenty people, including eight Polish nationals who hold executive positions.

Company activities

The company's core business remains unchanged and therefore it is still engaged primarily in the export of non-ferrous metal semi-fabricates from Poland to all destinations, but chiefly to Western Europe; and in the export of non-metal semi-fabricates to Poland. In addition, the company imports Polish bearings into the UK, although this still represents a small proportion of the company's business. It remains the case that the company will occasionally source products from outside Poland when the opportunity arises, and that Germany provides the largest market for the company's products.

It was felt that the benefits of a British location remained as relevant today as they did ten years ago, although the UK is not one of the principal markets for the company's product range. Location in London provides the company with advantages considered to be crucial in the international trading of metals, including accessibility to the major international metal traders, better market information, and access to the Metal Exchange, leading to rapid transactions at the operating market prices. These advantages have been complemented by the more personal contact the company has been able to establish with Polish producers and customers as a result of the recent improvements in telecommunications between Poland and the UK. The company also felt that it was now operating on a more profitable basis to that of ten years ago because of its years of experience in the market-place. It had established a name and reputation for itself, had developed many positive relationships with current and potential customers and was now able to raise necessary financial capital for various projects by virtue of its sound financial base.

One of the effects of the recent economic and political changes in Poland has been the reduced influence of foreign trade organizations in the conduct of foreign trade. Previously, Impexmetal held a state monopoly in the trading of metals, which it would import and export according to quotas decided by the Ministry of Planning, but now Polish factories have complete autonomy and can decide their own quotas and their own customers. Some of these factories have taken advantage of these new competences by attempting to export their own production, but their efforts were not completely successful partly because of their lack of experience and contacts. Consequently, most have reverted back to using a third party like Impexmetal to export their metal products. Although Impexmetal now has to compete with other companies for the purchase of Polish products, it has managed to retain much of its influence in Poland and remains financially secure, but it will undergo further change in 1993 when it is expected to be privatized. The company could not predict the effects of either this proposed privatization, or the reduced influence of foreign trade organizations in the conduct of trade, on its future mode of operation or its relationship with Impexmetal. It did, however, note that Impexmetal's ability to direct company policy may be declining. Whereas previously the company was concerned with fulfilling its obligations to Impexmetal and was regarded as the UK extension of Impexmetal's operations, it now has slightly more autonomy, and has become more commercially oriented in its dealings, carefully scrutinizing the profit margins on each transaction. The company still works closely with Impexmetal, however, and takes advantage of its representative bases in Germany, France, Belgium and Switzerland in its business transactions throughout Europe. The other foreign trade organization which founded the company and is a continuing shareholder is Anglo-Dal. It, however, has long ceased to play an active role in the operation of FLT and Metals Ltd., and the company is currently considering the possibility of buying its shareholding.

The company has not experienced any difficulties in securing supplies from Polish producers in recent years, although it has been affected by the rising costs of Polish supplies as spiralling inflation has forced production costs upwards. It was felt, however, that the company's main difficulty remained the same - finding customers willing and able to purchase its products.

Case study no.4 - Ridpath Pek Ltd

Company background

Ridpath Pek Limited, formed in 1976 from a previous agency relationship, is a jointly owned Anglo-Polish company primarily engaged in the distribution of Polish meat products in the UK. It has undergone changes both in terms of its ownership and the activities it performs over the course of the last ten years. Previously, Polish ownership of the company was divided between Animex, the Polish foreign trade organization responsible for the import and export of animal products; and Anglo-Dal, a British registered company then owned by the Polish foreign trade organization, Dal. The company remains 50-50 per cent Anglo-Polish owned, with the British owners still Ridpath Brothers Ltd, but Animex, a different legal entity now, having been privatized in the early 1980s, bought out Anglo-Dal's share in the company in 1984. This decision did not affect the company's mode of operation as Animex was, and continues to be, the company's principal supplier and therefore the active partner on the Polish side.

Company activities

Ridpath Pek Ltd. is still primarily involved in the distribution and sale of Polish meat products in the UK market. It sells to 350 customer outlets, although 70 per cent of its sales go to the top seven retailers in the country. It is responsible for undertaking all marketing and after-sales activities. However, during the last five years or so both the company's product range and the proportion of Polish products it imports have changed. In its early years, Polish bacon provided the company with its main source of income, and the company originally imported around 1000 tons of bacon per week. Bacon, however, was one of the products progressively affected by levies imposed as a result of the high level of protection accorded to EC members through the Common Agricultural Policy. This meant that bacon was no longer an economic commodity for the company to import from Poland or for the Polish authorities to export to the UK. A better return for Polish pig meat could be obtained from countries outside the EC, and, for example the Polish authorities concentrated on gaining a larger share of the US canned ham market. Consequently, the company now only imports around ten tons of bacon each week from Poland.

To compensate for the decline in bacon imports, Ridpath Pek was entrusted by Animex with the sale of Polish canned meats, although with reduced supplies since 1990. The company simultaneously encouraged Polish producers to move into new product areas such as stewed steak and minced beef. The latter two products were initially successful, but the flow of supplies of both products from Polish producers have not always been reliable. These difficulties have developed over the last few years as Polish factories have become more autonomous and the former central meat board ceased to exist. Animex now buys its products direct from the producing factories, and has to compete for supplies with customers in the home market. Whereas previously delivery of the products was virtually guaranteed because of the government imperative to earn hard currency, other pressures are now affecting supplies. For example, the convertibility of the zloty has meant that the prices that Animex can offer the producing factories are now affected by the zloty/US dollar exchange rate, which fluctuates considerably, having suffered two devaluations in 1991. The company has tried to overcome the problems of supply shortage by sourcing the same products from other EC countries to a strict specification under the Pek trademark, but this has only been achieved for canned meat products as the intrinsic qualities of Polish cooked meats make them difficult to replicate. The importing of similar products from other countries has meant that only 60 per cent of the current produce sold by the company is Polish, compared to over 90 per cent in the early 1980s.

In order to lessen its reliance on meat products, the company has also diversified into other product areas and attributes its continued existence to this factor. It now imports several non-meat products from Poland including jams, fruit, vegetables, a cereal beverage drink and confectionery, and these commodities accounted for 11 per cent of total company sales in 1991. Again, the company has been adversely affected by EC legislation, most notably when the dramatic increase in EC sugar levies in the late 1980s made the company's previously successful importing of Polish jam uneconomic. However, the company has greatly expanded its imports of Polish preserved fruit and vegetables (blackcurrants, bilberries, baby beets, sliced beetroot, sauerkraut and cucumbers), all of which have been successfully marketed and sold in the UK. The company launched its 'Barley Cup' beverage product in the early 1980s, which is now the market leader in its product range. 'Sesame Snaps' are the company's most successful product from the highly regarded Polish confectionery industry, and it is aiming to expand its range of imports from this sector in the future. The company also imports a few non-meat, non-Polish products such as herrings from Holland and confectionery from Austria.

As explained previously, the company felt that its business has been adversely affected by the growing adverse impact of the EC agricultural policy combined with rampant Polish price inflation. In an attempt to encourage relations in post-Communist conditions, the EC permitted the entry of Poland, Czechoslovakia and Hungary into the General System of Preferences in 1989, one of the effects of which was to concede under quota a 50 per cent cut in the levy affecting pork exports. While the company was grateful for this concession, it did not welcome certain aspects of the legislation which required each company to lay down a security of approximately £300 per ton for each quota applied for, to be fortfeited if the product was not imported. Although recognized as a necessary control mechanism, it was felt that this system penalized companies, like Ridpath Pek Ltd., who could not always guarantee the reliable delivery of supplies within a restricted time-scale. In 1992, this system was replaced by association agreements which have extended the concessions to a few more products, although the sum of money to be deposited remains the same. The company, therefore, remains dependent on factors outside its control - the stability and reliability of both Polish companies and the Polish economy. In practical terms, this uncertainty has forced the company to maintain closer contact with Polish producers resulting in frequent visits to Poland to assess the situation, as well as to continue its policy of diversifying its product range and supply sources.

Ridpath Pek Ltd. felt that a combination of the above factors were mainly responsible for slightly reduced earnings in recent years. However, the UK recession was also cited as a factor contributing to the company's diminished profitability in 1991, a year in which its annual turnover was £18 million, compared to some £26 million ten years ago at the peak of its bacon trade. As its products these days are at the high quality and more expensive end of the market they have also been more susceptible to the effects of this recession. Consequently, the company has diverted considerable resources towards increased advertising and marketing of its products in order to resist the loss of its market share. Overall, by reducing its costs and by careful diversification, it has continued to earn an adequate return on its capital.

The company was pleased that it had continued to operate satisfactorily amidst unfavourable economic conditions both in Poland and the UK. It remains optimistic about the future and is encouraged by economic indicators predicting lower inflation and stability in Poland as well as those forecasting an end to the UK recession.

Comments

The main commercial activities of the case studies in this sample consisted of the import into the UK of products made in the country of origin of the East European owners. As was the case ten years ago, the major exception to this general rule was FLT and Metals Company which remained primarily involved in world-wide export trading activities related to Polish products, using London as a location for reasons of access to the Metal Exchange. Ridpath Pek Ltd., however, had also been forced to source products from outside Poland in response to sporadic supply difficulties.

The companies were also all still engaged in export activities to the East European country in which the parent foreign trade organization was domiciled, although this remained a minor activity for Toolmex Polmach, FLT and Metals Company Ltd., and Ridpath Pek Ltd; and a less important activity for the London Chemical Company whose exports to Hungary had remained at the same level whereas the volume of Hungarian products that the company imported into the UK was £20 million in the early 1990s, compared to £4 million some ten years before.

As previously, the companies sometimes supplemented their import and export activities by importing from and exporting to other locations other than the country of ownership. This had always been a major activity of FLT and Metals Company, but both Ridpath Pek and Toolmex Polmach had found it necessary on occasion to diversify both in terms of product and market. As was the case ten years ago, none of the sample companies were involved in manufacturing activities. This could still be attributed to the lower direct costs of production in Eastern Europe, the apparent lack of many market niches in which the East European countries might have a technological advantage, and the predominantly 'trading' profile of the foreign trade organizations engaged in company ownership.

It is evident that there were many similarities between the commercial activities currently undertaken by these four companies compared to those undertaken ten years ago. It is also apparent, however, that the companies were operating amidst a background of turbulent economic and political change affecting both the parent foreign trade organization and the East European country concerned. The reality of this change was recognized by all four companies, but its actual effect on current operations varied from company to company. The London Chemical Company felt that it now possessed more autonomy, as evidenced by the utilization of its own account in each transaction. Furthermore, it now regards the main objective of the company as generating profit, whereas previously its primary aim was to provide a service to the Hungarian government. Similarly, Toolmex Polmach felt that it now operated as a company in its own right rather than as an extension of the foreign trade organizations' operations in the UK, and that it was consequently recognized as such. Although the Polish foreign trade organization, Impexmetal, has retained much of its influence in Poland and was still directing the company's commercial activities, FLT and Metals Company felt that it was slightly more autonomous than in the past and that it too had become more commercially oriented in its business transactions. Finally, Ridpath Pek Ltd. felt that a combination of external factors, most notably the European Community's CAP levies and poor economic conditions in Poland, had adversely affected the company's operation, and had forced it to diversify both product range and supply sources in order to remain competitive.

The companies felt, however, that it was too early to assess the effect of the recent changes in Eastern Europe on their operations, particularly as Chemol RT and Impexmetal are due to be privatized in the near future. All four companies did emphasize their continued profitability amidst unfavourable conditions and looked towards the future with optimism.

Notes

1. Hill, M.R. 'Soviet and East European company activity in the United Kingdom and Ireland' in Hamilton, G. (ed); *Red Multinationals or Red Herrings?,* Francis Pinter, London, 1986, pp. 17-87.
2. See McMillan, C.H., *Multinationals from the Second World*, Macmillan, London, 1986. The research carried out by McMillan and his team at Carleton University consisted of gathering of data on Soviet and East European investment on a global basis as presented in McMillan, C.H., Egyed, P. (eds), *East-West Business Directory : A Listing of Companies in the West with Soviet and East European Equity Participation*, Duncan Publishing, London, 1983. This directory was also updated in subsequent years, and the most recent version was published by the United Nations in 1992. (see Transnational Corporations and Management Division, Department of Economic and Social Development, United Nations Economic Commission for Europe, *The East-West Business Directory 1991/1992*, United Nations, New York, 1992). Other aspects of the Carleton group's research relating to inward and outward investment are included in Transnational Corporations and Management Division, Department of Economic and Social Development, United Nations Economic Commission for Europe, *World Investment Directory 1992 : Volume 2, Central and Eastern Europe,* United Nations, New York, 1992.

8 Comments, conclusions and further research

Comments

The research described in this book has studied the experiences of a sample of British companies engaged in commercial relations with the former socialist countries of Eastern Europe. The selected sample included some forty British companies previously engaged in a range of trade, industrial co-operation and technology transfer arrangements with the USSR and Eastern Europe during the late 1960s, the 1970s and the early 1980s.[1,2]

The companies in that group were drawn from a wide range of engineering industries, and were considered by the present authors to be representative British companies having established experience in trading with the former socialist countries. Furthermore, it was considered that re-interviewing of these companies provided the opportunity to carry out a longitudinal study of their experiences extending from the early 1980s to the early 1990s, a decade in which there have been major political changes in the former Soviet Union and Eastern Europe.

The information obtained from the case studies in chapter 2 suggests a sharp decline in the export of capital equipment by British companies to the former socialist countries of Eastern Europe, compared with the levels of export deliveries some ten to fifteen years previously. To a great extent, this has been caused by major reductions in demand from customers in the former USSR, who had previously been important buyers. The major reason advanced for these changes in the level of demand was an acute shortage of hard currency by buyers in the CIS, with this shortage more apparent in the individual factories now responsible for their own foreign trading activity. These decreases in demand, combined with concerns over factory creditworthiness in the former Soviet Union, have led some UK companies to re-organize their export policies and practices from selling direct to the customer, to selling through Western third parties.

It is apparent, however, that these conclusions were reached from the study of a very small sample, in which several of the companies sold their products to the Soviet automotive industry, a sector which experienced an investment boom in the late 1960s and early 1970s, but which had not been maintained during the 1980s. Furthermore, this study was carried out against a background of rapid contraction in the British machine tool industry, in which declining market activity in the socialist countries could be caused by some companies ceasing to trade altogether as well as other companies being unsuccessful in the CIS and East European markets. One company, though, was engaged in technology transfer discussions with a large Russian truck plant, whilst the remaining companies were hopeful of future business possibilities in that region.

Some of the material in chapter 3 reinforces some of the conclusions presented in chapter 2. In the first place, those companies which had originally sold licences in the 1960s and 1970s for technology related to the automotive and agricultural engineering industries in Poland and Romania, did not receive much follow-on business related to those commercial arrangements, during the 1980s and 1990s. As in the case of Soviet investment in the automotive industries discussed in chapter 2, therefore, the East European investment in automobile production and agricultural engineering during the 1960s and 1970s appears to have been one-off large scale funding to increase capacity in those sectors. A further similarity with some of the companies reported in chapter 2 is also revealed by the radical changes taking place within some of those companies, affecting their organizational structure or ownership during the 1980s and 1990s, which in turn affected their trading links with the former USSR and Eastern Europe.

Of the ten companies described in chapter 3 above, six have maintained their trading links with the region. The experiences of some of these firms in this market have been far from negative, and point to positive signs for success in the new transitional conditions operating in the CIS and East European markets. One factor influencing success has been the capability of being able to sell to CIS factories having supplies of foreign currency, who also need technology and know-how to reorientate their product mix from military to civilian items. In other cases, companies appear to have been successful when they have been able to relate their sales to purchases of CIS products for which there appears to have been definite market potential in the West, or where there might be some savings in foreign currency expenditure. These proved to be related to natural resource extraction industries or industrial materials, or manufactured products in which CIS expertise could be adapted for application in Western markets in a cost effective fashion. The realization of these potentials has required the outlay of high levels of corporate resource by the successful British firms, and the ability to develop and construct the appropriate financial frameworks to support the required directions of business, sometimes achieved through joint ventures.

This requirement in investment in corporate resources, has also been seen in the two cases of large scale industrial co-operation presented in chapter 4, as a consequence of the high levels of corporate skill required to develop and implement such complex arrangements. A major factor emerging from the two case studies in chapter 4, however, is the high degree of risk existing in this type of large scale transaction if internal conditions in the East European partner's country fail to support the intended business development. It is apparent, therefore, that there are still many lessons to be learned from these case studies which will have some relevance to companies likely to get more involved in manufacturing in the former Soviet Union and Eastern Europe.

Of the five companies who had originally purchased licences from the USSR and Eastern Europe described in chapter 5, only one has continued to use the process originally purchased during the 1960s and 1970s, and that process is only used very rarely. The other four companies have concluded their licensing work, and three of these have neither maintained nor expanded their links with the former Soviet Union and Eastern Europe. This suggests that the purchase of licences from the former USSR and Eastern Europe although always small, is now almost nonexistent.

The exception to this general trend towards severing of trade links with the former socialist countries described in chapter 5, is a company forming part of a large engineering group, which has radically expanded its activities in the region, mainly through the efforts of its recent acquisitions from the UK steel industry. The flow of technology has been exclusively in an eastward direction, however, illustrating the demise of the westward flow of technology from the former USSR and Eastern Europe This company has been able to capitalize on the demand for steel in the former Soviet Union, the requirement to update the technology of existing and new capacity, the availability of existing natural resources of ores and other materials, and the potential for the sale of steel in the West on a price effective basis. The development of these capabilities, as in the case of those successful companies described in chapters 3 and 4 above has required a high level of investment in corporate resources, and the ability to develop a financial framework to support the business operation.

Of the five companies previously involved in technical co-operation with the former socialist countries of Eastern Europe, described in chapter 6, only one company is still involved in technical co-operation with the same partner. Although internal organizational changes were the reasons for the decline of technical co-operation in two of the cases, it is probable that this type of agreement will tend to become outmoded as decentralization continues in the former socialist countries, and Western-style business practices for technical collaboration are used more widely.

The four companies interviewed in chapter 7, which were either wholly-owned or partly-owned by organizations based in the former socialist countries of Eastern Europe, illustrated that their location in the UK could continue to be used advantageously to export East European products to the West, and to engage in other profitable business activities. It appears likely, therefore, that investment in these companies will continue, although on a level consistent with the availability of scarce foreign currency. It became apparent during this research, however, that some of these companies were having to compete more rigorously to secure adequate supplies from within the socialist countries, particularly as links were being severed between factories and foreign trade organizations. Suppliers can therefore find alternative buyers on the newly established and more open markets, rather than having to sell through defined distribution channels as previously.

The apparent declining levels of British export activity in the former socialist countries of Eastern Europe, suggested from the results in chapters

2, 3 and 4 are supported by a study of international trade statistics for the 1980s. OECD statistics for total UK exports to the former socialist countries of Eastern Europe indicate a decline in exports in dollar current prices in 1981, 1982 and 1983 compared with a peak year in 1980, although there were signs of improvement during the mid to late 1980s compared with the mid-year of that decade.[3] The volume of UK exports to the region were probably static in real terms and physical quantities, however, when the effects of inflation and movements in the exchange rate of sterling against the dollar are taken into account,[4] and British exports continued to decline in competitiveness in that region,[5] as explained in the following paragraph. This loss of competitiveness is particularly significant in the context of this current research, as it covers the time period when companies would need to modify their market priorities and production strategies in line with achieved sales performance as well as anticipated demands.

The parameter selected to define competitiveness has been a particular country's share of total exports by the OECD countries to the socialist countries of Eastern Europe. Although there will be variations from year to year caused by exchange rate movements and some random factors, a comparison of these shares provides a basis for analysis. Although there were some variations from year to year, the share of UK exports within the overall total of OECD exports to the former socialist countries of Eastern Europe, was in overall decline from some 5.7 per cent in 1980 to 4.5 per cent in 1990, whilst Japan, although apparently declining in terms of market share between 1987 and 1990, had market shares varying between 6.6 per cent and 10.7 per cent over the decade, which were greater than those of the UK. Western Germany in particular increased its share in that region, which varied between 19.3 per cent and almost 29 per cent.[6] The decline in British market share is seen more clearly, however for those products which have been major product segments for the Comecon countries, namely 'manufactured goods classified by material', and 'machinery and transport equipment', which accounted for some 25 per cent and 30 per cent respectively, of OECD exports to that region during the mid 1980s. The UK market share declined from 5.3 per cent in 1980 to 2.6 per cent in 1990 for manufactured goods classified by material, and from 8.1 per cent to 3.9 per cent for machinery and transport equipment.[7]

To a certain extent, however, this performance is a reflection of British manufacturing industry's international competitiveness in general, as the UK share of OECD countries' world exports fell from 9.3 per cent in 1980 to 7.7 per cent in 1990, but the reduction in market share was slightly more acute for materials classified by material (9.5 per cent in 1980 to 7.0 per cent in 1990) and approximately the same for machinery and transport equipment (9.1 per cent in 1980 to 6.7 per cent in 1987).[8] These figures also show that British industry has achieved a lower market share in the former socialist countries compared to its market share for exports to the rest of the world, although similar patterns are also apparent for Japan and the USA, but not for Germany, Italy, Austria and Finland whose market shares in the socialist countries have been far higher than in the rest of the world, particularly for manufactured goods and engineering products.[9]

In spite of this general decline in market share for British goods, however, there have been some particular product segments where British market shares have been comparatively high, and to a certain extent these statistics reflect the corporate marketing effort expended by companies described in this book. In the case of Poland, for example, the share of OECD exports of machinery and transport equipment accounted for by the UK, although in decline has varied between 15.7 per cent and 6.9 per cent between 1980 and 1987. In the case of Romania, the competitiveness of British machinery and transport equipment is even more apparent, as the share of OECD exports of these products originating from the UK accounted for 10.9 per cent in 1980, before increasing to a peak of 33 per cent in 1983, and achieving a high level of almost 22 per cent in 1987.[10] To a great extent, these successes in market share are due to the efforts of Massey-Ferguson-Perkins to establish an industrial co-operation agreement with a foreign trade organization in Poland, and of British Aerospace to organize an industrial co-operation agreement in Romania.

In addition, the share of OECD exports of power engineering equipment delivered to the USSR, accounted for by British companies achieved 23.9 per cent, 12.7 per cent and 13.6 per cent in 1982, 1983 and 1984 respectively, whilst the UK share of OECD world exports of power engineering equipment accounted for some 14.7 per cent, 11.7 per cent and 10.7 per cent over the same time interval.[11] This high market share for power generation equipment was mainly due to the successful export activity of one UK power engineering company, but as sales decreased during the late 1980s because of scarce foreign currency and domestic substitution by the Soviet buyer, the company's parent group has had to consider other product markets and methods of payment (see case study no.3, chapter 3).

Conclusions

The major conclusion emerging from this research of a sample of British companies is that there has been a significant decline in the commercial activities of UK companies with the former socialist countries of Eastern Europe, compared to the levels of commercial activity encountered some ten to twelve years ago. The predominant reason given for this decline of commercial activity has been the reduced market opportunities in the former Soviet Union and Eastern Europe as the foreign currency purchasing power of these countries has declined. This has been caused partly by declining world energy prices in the 1980s which led to reduced foreign currency earnings for the USSR, and the low level of competitiveness of Soviet and East European manufactured goods in world markets, particularly in terms of non-price characteristics. Each of these factors have become aggravated by scarce foreign currency being used to purchase items previously traded for transferable roubles through the Council for Mutual Economic Assistance.[12]

To a certain extent, this decline in foreign currency purchasing power has also been influenced by the decentralization of export and import responsibility from centralized foreign trade organizations to individual factories and other institutions. These production, distribution and research organizations have normally required smaller purchases than those previously ordered by the foreign trade enterprises for large centrally sanctioned projects. The limited purchasing power of these decentralized organizations has been further decreased by their inability to export significant production volumes to cover import purchases, and the withdrawal of cross-subsidies available through the previous central planning system. The final factor reducing purchases by these organizations has been the rapidly deteriorating exchange rate for the currencies of the former Soviet Union and Eastern Europe, combined with increasing taxes to be paid from export earnings.[13]

Decentralization of foreign trade in the former Soviet Union and Eastern Europe, although having the beneficial effect of linking end-users much closer to foreign suppliers, has also raised questions about the creditworthiness of many of the producers now having foreign trade rights. This has consequently given rise to a growing sense of caution amongst British exporters, reflected in the reduced willingness of banks to lend for projects in the former socialist countries, and a lack of official government support, until recent months, through the Export Credit Guarantee Department.

The second set of factors influencing British companies' perceptions of market possibilities in the previous socialist countries of Eastern Europe have been the general political and economic uncertainties in that region. These perceptions contrast with the euphoria surrounding the ending of communist one-party power in the Soviet Union and Eastern Europe, which was originally welcomed by the majority of Western government and business leaders as a consequence of the anticipated reductions in international tension, and potential opportunities for business in that region during the transformation to a market economy. In practice, however, the process of transformation has been accompanied by high inflation, industrial dislocation, and rapid and unpredictable political change as evidenced by the attempted coup in the USSR in August 1991, followed by the fragmentation of the Soviet Union into separate republics. The former Soviet Union and other Eastern Europe countries have repeatedly requested aid from the Western economies for humanitarian purposes, technical assistance, currency support and industrial investment. To a great extent, therefore, future possibilities for exports to that region will probably be influenced by the level of financial assistance from the Western economies to support these activities.

The third set of factors influencing the current level of commercial activity in the former socialist countries of Eastern Europe, compared with some ten years ago, would appear to be the changes in investment strategy and buyer preferences of those Soviet and Eastern European customers. Previously, the former socialist countries were engaged in investment programmes related to rapid expansion and refurbishment of the automotive and agricultural engineering sectors, and many of their imports from the sample of British companies were connected to those programmes. This rapid expansion was achieved through the building of new factories and the renovation of existing facilities, and represented a major one-off investment decision. As some of the industrial co-operation agreements with the sample of British companies were linked to licensing arrangements, it was found that these agreements lapsed as the relevant licence expired, and were not subsequently extended.

A fourth major set of factors affecting the present activity of British companies in the former socialist countries of Eastern Europe has been the macro-economic environment in which British companies have been operating over the recent decade. The dual effects of high interest rates and exchange rates on the UK's manufacturing industry have been documented elsewhere,[14] but the major consequences have been company contraction and liquidation. Most of the companies covered in this survey were facing problems of declining international and domestic markets, thereby leading to policies of rationalization and contraction, or changes in ownership or liquidation. The general effect of these changes in ownership and management policy has been withdrawal from the former Soviet and Eastern European markets, particularly where high levels of promotional investment were required to retain a market position in that region.

There are several companies, however, who have become more actively involved in the former Soviet and East European markets, through the vehicle of joint ventures. The major advantages perceived by investment in these joint ventures were the possibilities for future business in that region, particularly for companies involved in business related to natural resource extraction and heavy engineering. Part ownership in a joint venture enabled the British company to use that vehicle for business relating to new construction and refurbishment in the USSR, as a consequence of the well established presence in that market and expertise of the Soviet partner, combined with British technological expertise and experience in international marketing of products related to the project investment. A further powerful reason for investment and participation in joint ventures has been that of the adoption of a defensive strategy as many European, American and Japanese competitors also became joint venture partners with Soviet organizations.

Further research

Each of the British partners in the above-mentioned joint ventures had many years of prior experience in the Soviet and East European markets, and their progress in these markets through joint ventures is clearly worthy of future research. Research on non-joint venture companies should also be maintained, however, particularly those companies entering the market for the first time, and especially those with expertise in natural resource extraction as the former USSR is still dependent on such products to expand its exports. Many joint ventures involving other British companies have also been established, some with little previous experience of marketing to the socialist countries and the progress of these companies should consequently be monitored.

Research on a sample of eight Western companies (three British, three German and two Austrian) has recently been carried out[15] using methodology similar to that used in the research described in this book, namely structured interviews. The research extended from 1986 to 1988, with an update in 1991. Some of the objectives of that study were different to this present research, such as the type of joint venture model chosen by the partners and the amount of statutory capital invested by the Western partner; but other objectives were very similar, relating to the motivations for both sides entering the agreement as perceived by the Western company, and the extent of the Western partners' business experiences in the former socialist countries of Eastern Europe before participating in the joint venture. All of the joint ventures in the sample were engaged in manufacturing activities, and all but one of the companies interviewed, had been trading with the Comecon countries for at least twenty years, having previous experience in industrial co-operation with their joint venture partners. All of the companies were primarily interested in entering a joint venture in order to secure their market position in the former socialist countries, although one company saw joint venture arrangements as a means to reduce countertrade deals, and others saw the potential for using a lower cost production base for certain goods destined for Western markets. The Eastern partners, in their turn, were mainly interested in gaining access to Western technology and know-how, and substituting hard currency imports through the joint venture with possibilities for access to hard currency markets through their Western partners' distribution outlets.

The Western partners' contributions to support those objectives consisted primarily of know-how and technology provided mainly through the sale of licences and the training of technical staff, and assistance in the implementation of this technology by the retention of overall responsibility for quality control. The Eastern partners in their turn provided factory premises and direct labour, and local market and supplier knowledge.

The general view of these Western companies was one of satisfaction with the arrangements, and the majority of them increased their equity shareholding during the time of the research. Since 1991, however, some companies in the sample were reconsidering their investment strategy because of the political and economic uncertainties in the region, particularly in the former USSR and Bulgaria where provisions for profit transfers and quality control apparently continue to cause problems for some investors. In addition, some companies were far from certain as to how privatization within the former socialist countries of Eastern Europe would affect their investment strategies in that region. One particular cause of concern was that competitors would be able to obtain an advantage through purchase of the assets of the Eastern partner in the joint venture.

Following the rapid growth of joint venture arrangements in the USSR after changes in Soviet joint venture legislation in 1989, a postal survey of Anglo-Soviet joint ventures was also carried out in 1990 and 1991.[16] A postal questionnaire method was chosen in order to obtain information in a relatively short space of time from quite a large sample, and the questions selected related to similar topics covered in the previous survey.

Of the twenty nine companies contacted, only nine replies were received, and one of those claimed never to have been engaged in a joint venture in the former USSR. The eight companies participating in the survey were engaged in service, manufacturing and construction activities, with average capital investments in the USSR amounting to just over £1 million. A considerable proportion of the sample had very little business experience in the former USSR before entering into the joint venture arrangement. This may explain the generally high level of disappointment experienced by these firms in some of the joint venture arrangements and in the general business environment, as these companies would have had no significant experience of business conditions in the former USSR or the time required to establish mutually profitable commercial relations. The disappointment of these companies was compounded, however, by difficulties in the repatriation of profits in hard currency.

It is clearly important, therefore to continue to study the experiences of joint venture companies to assess how they continue to trade in the rapidly changing political and economic environment in the former socialist countries of Eastern Europe, and the opportunities for privatization continue.

Finally, the experience of 'competitor' West European, Japanese and American companies should be studied in further detail, particularly as Western technical and financial assistance for the former socialist countries may begin to increase. This is particularly important at present, as British manufacturing industry as a whole apparently continues to lose market share in the former USSR and Eastern Europe, mainly to German and Japanese competitors.[17] Although the market is currently contracting in total,[18] a continued loss of market share could make it more difficult for British firms to become more active in that region if trading conditions begin to improve.

Notes

1. Hill, M.R., *The Export Marketing of Capital Goods to the Socialist Countries of Eastern Europe*, Wilton House, Farnborough, 1978.
2. Hill, M.R., *East-West Trade, Industrial Co-operation and Technology Transfer*, Gower, Aldershot, 1983.
3. *OECD Foreign Trade Statistics*, Series C provide the following data for British exports to the European members of Comecon. Data available in the 1987 and 1990 editions have been used.

1980	$2.43 bn
1981	$1.94 bn
1982	$1.48 bn
1983	$1.42 bn
1984	$1.73 bn
1985	$1.52 bn
1986	$1.71 bn
1987	$1.82 bn
1988	$2.05 bn
1989	$2.21 bn
1990	$2.25 bn

4. It is difficult to apply an inflation factor for export sales, because of the influence of international market conditions as well as domestic inflation. If we look at the latter factor alone, however, the producer price index of UK manufacturing industries, other than food, drink and tobacco increased from an index of 100 in 1985 to 120 in 1989, and the value of the 1985 pound, based on the retail price index was 82 pence in 1989. (See *Annual Abstract of Statistics 1991*, HMSO, London, 1991, Table 18.2 (Producer price index numbers of output (home sales) and Table 18.6 (Internal purchasing power of the pound, based on RPI).

The conversion rate for the pound against the dollar used in OECD statistics is provided in OECD *Foreign Trade Statistics Series C*, 1990, Vol.2 for the years 1985-90, and this index has been used to convert the OECD dollar values to sterling values, and then apply a price deflator as shown in the table below.

Year	UK exports to COMECON ($) see fn3	Dollar/Sterling exchange rate ($)	UK exports to COMECON (£)	Producer Price Index	UK exports to COMECON in 1985 prices
1985	$1.52 bn	$1.29	£1.18 bn	100	£1.18 bn
1986	$1.71 bn	$1.46	£1.17 bn	104.1	£1.12 bn
1987	$1.82 bn	$1.64	£1.11 bn	108.6	£1.02 bn
1988	$2.05 bn	$1.78	£1.15 bn	113.8	£1.01 bn
1989	$2.21 bn	$1.63	£1.36 bn	120.0	£1.13 bn

5. British exports to the region were also losing market share in the 1970s. (see Hill, 1983, pp.13-48).

6. The share of OECD exports to the socialist countries of Eastern Europe during 1980-90 were as follows:

Year	UK	FRG	JAPAN
1980	5.7%	22.3%	8.5%
1981	4.9%	19.3%	10.2%
1982	4.9%	19.3%	10.2%
1983	4.1%	22.1%	10.2%
1984	5.1%	21.0%	8.9%
1985	4.5%	21.6%	9.8%
1986	4.8%	25.2%	10.7%
1987	4.8%	26.1%	8.6%
1988	4.7%	25.5%	8.9%
1989	4.4%	26.1%	7.5%
1990	4.5%	28.7%	6.6%

7. The UK shares of OECD exports of Manufactured Products Classified by Material, and Machinery and Transport, to the socialist countries of Eastern Europe, have been as follows:

Year	Manufactured Products Classified by Material	Machinery and Transport Equipment
1980	5.3%	8.1%
1981	3.9%	5.6%
1982	2.6%	5.2%
1983	2.7%	4.7%
1984	5.5%	5.8%
1985	3.4%	3.8%
1986	2.8%	3.6%
1987	3.0%	3.8%
1988	2.9%	4.2%
1989	2.7%	4.1%
1990	2.6%	3.9%

8. The UK share of OECD world exports were as follows:

Year	Total Exports	Manufactured Products Classified by Material	Machinery and Transport Equipment
1980	9.3%	9.5%	9.1%
1981	8.2%	6.7%	7.4%
1982	8.4%	6.7%	7.4%
1983	8.0%	6.7%	6.5%
1984	7.8%	6.5%	6.2%
1985	8.0%	6.4%	6.5%
1986	7.3%	6.5%	6.2%
1987	7.6%	6.8%	6.7%
1988	7.4%	6.8%	7.0%
1989	7.3%	6.6%	7.1%
1990	7.7%	7.0%	7.4%

9. In 1985, for example, the proportion of OECD exports to the world and the socialist countries of Eastern Europe were as follows:

Country	% of OECD world exports			% of OECD exports to the socialist countries of Eastern Europe		
	Total	Manufactures	Machinery	Total	Manufactures	Machinery
UK	8.1%	6.4%	6.5%	4.5%	3.4%	3.8%
Japan	14.0%	13.7%	24.1%	9.8%	14.8%	14.4%
USA	16.4%	6.8%	19.1%	9.4%	0.4%	2.8%
Germany	14.6%	16.3%	17.1%	21.6%	26.4%	29.4%
Italy	6.3%	8.5%	5.0%	7.7%	12.3%	9.0%
Austria	1.4%	2.8%	1.1%	5.6%	9.0%	5.0%
Finland	1.1%	2.5%	0.7%	9.3%	9.7%	15.6%

Compiled from Hughes, J. 'A Statistical Review of Western Exports to Eastern Europe, and the Effects of Export Controls on East-West Trade' MBA Dissertation, Loughborough University, 1991, Vol.2, pp. 118, 126, 127.

10. See Hughes (1991)
11. Hill, M.R., 'Les relations économiques URSS-Royaume-Uni'; *Le courrier des pays de l'est*, no. 354 - Novembre 1990; pp. 33-56.

12. Figures from *Ekonomicheskaya Gazeta* no. 13, March 1992, published in the *Moscow Narodny Bank Press Bulletin*, no. 1055, 15 April 1992, pp. 15, 18, illustrate the decline in the USSR's trade with the other former socialist countries in 1991, compared to 1990. Although Soviet imports in total in 1991 were only 57.6 per cent of the level they were in 1990, it was the imports from the former CMEA which suffered the greatest reduction, contracting to 38.7 per cent of their 1990 level in 1991. Imports from the industrially developed countries declined to 69 per cent of their 1990 level in 1991.
13. Hill, M.R 'Technology and Management in the Commonwealth of Independent States' in *European Business Review*, vol. 92, no. 4 (1992), pp. 3-15.
14. For a concise summary of these points, see Keegan, W., *Mrs Thatcher's Economic Experiment*; Allen Lane; London 1984, and Keegan, W.; *Britain Without Oil*; Penguin; London; 1985. Similar themes related to UK manufacturing decline in the 1980s are discussed in Newton, S., Porter, D., *Modernization Frustrated : The Politics of Industrial Decline in Britain since 1900*, Unwin Hyman London, 1988, pp. 185-210; Sharp, M., Shepherd, G., Marsden, D., *Managing Change in British Industry*, International Labour Office, Geneva, 1987, pp. 11-38; Judge, A., Dickson, T., 'The British State, Governments and Manufacturing Decline' in Judge, D., Dickson, T., (eds) *The Politics of Industrial Closure*, Macmillan, London, 1987, pp. 1-34. The view that surviving UK manufacturing companies are now better managed than in previous decades can be found in Bowen, D., *Shaking the Iron Universe : British Industry in the 1980s*, Hodder and Stoughton, London, 1990.
15. Watkins-Mathys, L.L.N.; 'The Role of Joint Ventures in East-West Trade, up to 1990', unpublished Ph.D. thesis, Loughborough University, 1992.
16. *Ibid.*
17. Figures from *Ekonomicheskaya Gazeta* no. 13, March 1992 published in the *Moscow Narodny Bank Press Bulletin*, no. 1055, 15 April 1992, p. 19, show that Soviet imports from the UK in 1991 reached only 44.2 per cent of their level in 1990 - 1,200.5 mn. roubles (commercial rate) in 1991, compared to 2,717.7 mn. roubles in 1990. This is significantly lower than the UK's industrial competitors eg. the level of exports from the FRG to the former USSR in 1991 was 61.6 per cent of the level attained in 1990. The corresponding figure for France was 78.7 per cent, Japan 70.7 per cent, Italy 89.7 per cent and the USA 108.1 per cent.
18. Figures from *Ekonomicheskaya Gazeta* no. 13, March 1992 published in the *Moscow Narodny Bank Press Bulletin*, no. 1055, 15th April 1992, p. 15 show that the former USSR's foreign trade reduced from 275,466.7 mn. roubles (commercial rate) in 1990, to 160, 649.0 mn. roubles in 1991; and that its trade with the industrialized West had declined from 121,615.3 mn. roubles to 92,066.6 mn. roubles over the same period.

Appendix

Topics for Structured Interviews at Companies Described in Chapter 2

COMPANY BACKGROUND

- Ownership
- Products
- Markets

PREVIOUS TRADING LINKS

- Products
- Markets

CURRENT TRADING LINKS

- Products
- Markets

FUTURE PLANS

Topics for Structured Interviews at Companies Described in Chapter 3

COMPANY BACKGROUND

Ownership
Products
Markets

TRADING LINKS

Changes in Company Policies and Ownership
New Agreements

INDUSTRIAL CO-OPERATION AGREEMENTS

Partners
Activities
Duration
Earlier Business
Current Status
Partners
Range of Activities

MOTIVATION FOR COMPANY TO ENTER AGREEMENTS

MOTIVATIONS FOR SOVIET OR EAST EUROPEAN PARTNER TO ENTER AGREEMENT

FINANCIAL ASPECTS OF AGREEMENTS

FUTURE PLANS

Topics for Structured Interviews at Companies Described in Chapter 4

COMPANY BACKGROUND

Ownership
Products
Markets

INDUSTRIAL CO-OPERATION AGREEMENT

Current Status
Previous Development
Future Plans

FINANCE

Initial investments
Volume of Countertrade
Mutual obligations

NEW TRADING VENTURES

Topics for Structured Interviews at Companies Described in Chapter 5

NUMBER OF LICENCE PURCHASES

PARTNERS

TECHNICAL FEATURES

DURATION

PREVIOUS BUSINESS ARRANGEMENTS

LICENCE CONTENTS

Licence purchase
Technical assistance
Manufacturing Equipment
Components and Materials
Quality Control
Training
Exchange of Information
Marketing

COMPANY MOTIVES FOR LICENCE PURCHASES

MOTIVES OF SOVIET OR EAST EUROPEAN PARTNER FOR SALE OF LICENCES

CONFORMANCE WITH INTERNATIONAL PROCEDURES

Topics for Structured Interviews at Companies Described in Chapter 6

NUMBER OF AGREEMENTS

PARTNERS

TECHNICAL FEATURES OF THE AGREEMENTS

DURATION

PREVIOUS BUSINESS ACTIVITIES

ACTIVITIES

- Information Exchange
- Co-ordination of research and development
- Training
- Licences
- Components
- Manufacturing Equipment

COMPANY MOTIVES FOR AGREEMENTS

MOTIVES OF SOVIET OR EAST EUROPEAN PARTNER FOR AGREEMENTS

Topics for Structured Interviews at Companies Described in Chapter 7

COMPANY BACKGROUND

Ownership
Personnel
Agencies

COMPANY ACTIVITIES

Objectives
Products
Products
Markets
Distribution

COMPANY OPERATIONS

Changes in Eastern Europe and former USSR affecting

(i) Company
(ii) Owners

COMPANY PERFORMANCE

Income
Profits
Business Plans

Bibliography

Amann, R., Cooper, J.M., Davies, R.W. (eds); *The Technological Level of Soviet Industry*, Yale University Press, New Haven, Conn. and London, 1977.

Aylen, J., Neglected Steel - The Future of Steel in Eastern Europe and Developing Countries - unpublished report, 1990.

Batt, J., 'The Political Context' in Rollo et al (1990), pp.6-39.

Bowen, D., *Shaking the Iron Universe : British Industry in the 1980s*, Hodder and Stoughton, London, 1990.

Clague, C., Rausser, G.C. (eds), *The Emergence of Market Economies in Eastern Europe*, Blackwell, Oxford, 1992.

Cooper, J.M. 'Iron and Steel' in Amann, Cooper and Davies (1977), pp.83.-118.

Ellman, M., Kontorovich, V., *The Disintegration of the Soviet Economic System*, Routledge, London, 1992.

Gardner, C., *British Aircraft Industry,* Batsford, London, 1981.

Goldman, M.I., *What Went Wrong with Perestroika*, W.W. Norton, London and New York, 1992.

Hamilton, G. (ed) *Red Multinationals or Red Herrings? The activities of enterprises from socialist countries in the West*, Francis Pinter, London, 1986.

Hayward, K., *The British Aircraft Industry*, Manchester University Press, Manchester, 1989.

Hewett, E.A., *Reforming the Soviet Economy*, The Brookings Institution, Washington, DC, 1988.

Hill, M.R., *Export Marketing of Capital Goods to the Socialist Countries of Eastern Europe*, Wilton Publications, Farnborough, 1978.

Hill, M.R., *East-West Trade, Industrial Co-operation and Technology Transfer*, Gower, Aldershot, 1983.

Hill, M.R., 'Soviet and East European company activity in the United Kingdom and Ireland', in Hamilton, (1986), pp.17-87.

Hill, M.R., 'Les relations economiques URSS-Royanne Uni' in *Le courrier des pays de l'est*, November, 1990, no 354, pp.33-56.

Hill, M.R. *Soviet Advanced Manufacturing Technology and Western Export Controls*, Avebury, Aldershot, 1991.

Hill, M.R., 'Management in the Commonwealth of Independent States', *European Business Review*, 1992, Vol,92, No.4, pp.3-15.

Hughes, J., 'A Statistical Review of Western Exports to Eastern Europe and the Effects of Export Controls on East-West Trade', unpublished MBA Dissertation, Loughborough University, 1991.

Judge, D., Dickson, T. 'The British State, Governments and Manufacturing Decline' in Judge and Dickson (1987), pp.1-34.

Judge, D., Dickson, T., (eds) *The Politics of Industrial Closure*, Macmillan, London, 1987.

Keegan, W., *Mrs. Thatcher's Economic Experiment*, Allen Lane, London, 1984.

Keegan, W., *Britain Without Oil*, Penguin, London, 1985.

Malcolm, N., 'How does the Soviet Union fit in' in Rollo et al (1990), pp.83-99.

McMillan, C.H., *Multinationals from the Second World*, Macmillan, London, 1986.

McMillan, C.H., Egyed, P. (eds), *East-West Business Directory : A Listing of Companies in the West with Soviet and East European Equity Participation*, Duncan Publishing, London, 1983.

Newton, S., Porter, D., *Modernization Frustrated : The Politics of Industrial Decline in Britain since 1900,* Unwin Hyman, London, 1988.

Rollo, J.M.C., 'Western policy : the room for manoeuvre' in Rollo et al (1990), pp.100-15.

Rollo, J.M.C., Granville, B., 'The agenda for reform' in Rollo et al (1990), pp.63-82.

Rollo, J.M.C. et al, *The New Eastern Rurope : Western Responses*, Francis Pinter, London, 1990.

Sharp, M., Shepherd, G., Marsden, D., *Managing Change in British Industry,* International Labour Office, Geneva, 1987.

Thomas, S., 'The Political Economy of Privatization, Poland, Hungary and Czechoslovakia' in Clague and Rausser (1992), pp.279-95.

Transnational Corporations and Management Division, Department of Economic and Social Development, United Nations Economic Commission for Europe, *The East-West Business Directory 1991/1992*, United Nations, New York, 1992.

Transnational Corporations and Management Division, Department of Economic and Social Development, United Nations Economic Commission for Europe, *World Investment Directory 1992 : Volume 2, Central and Eastern Europe*, United Nations, New York, 1992.

Watkins-Mathys, L.L.N., 'The Role of Joint Ventures in East-West Trade up to 1990', unpublished Ph.D. thesis, Loughborough University, 1992.

Index